Power Electronics and Power Systems

The Series, *Power Electronics and Power Systems*, encompasses most areas of power electronics, electric power restructuring, and power systems in general. It focuses on publishing advanced level textbooks, state-of-the-art titles, research monographs, professional books, and reference works related to the areas of electric power transmission and distribution, energy markets and regulation, electronic devices, electric machines and drives, computational techniques, and power converters and inverters. The Series publishes both authored books and edited compilations.

All titles are peer reviewed prior to publication to ensure the highest quality content. The Series features the leading international scholars and researchers as authors.

M.A. Pai, Professor Emeritus, Dept. of Electrical and Computer Engineering, University of Illinois, Urbana Champaign, IL 61801, 217-333-6790 (O), 217-344-0977 (R)

Alex M. Stankovic, Professor, Dept. of Electrical & Computer Eng., 440DA, Northeastern University, 360 Huntington Ave., Boston, MA 02115, (617) 373-3007

More information about this series at http://www.springer.com/series/6403

Mladen Kezunovic • Sakis Meliopoulos
Vaithianathan Venkatasubramanian • Vijay Vittal

Application of Time-Synchronized Measurements in Power System Transmission Networks

Springer

Mladen Kezunovic
Department of ECEN
Texas A&M University
College Station, TX, USA

Vaithianathan Venkatasubramanian
School of EECS
Washington State University
Pullman, WA, USA

Sakis Meliopoulos
School of ECE
Georgia Tech
Atlanta, GA, USA

Vijay Vittal
School of Electrical
Computer and Energy Engineering Ira
A. Fulton Schools of Engineering
Arizona State University
Tempe, AZ, USA

ISSN 2196-3185 ISSN 2196-3193 (electronic)
ISBN 978-3-319-34993-0 ISBN 978-3-319-06218-1 (eBook)
DOI 10.1007/978-3-319-06218-1
Springer Cham Heidelberg New York Dordrecht London

Printed on acid-free paper

Springer is part of Springer Science+Business Media (www.springer.com)

Preface

This book has been created through a research project titled "Future Grid Initiative" funded by the Department of Energy under a grant to the NSF Power Systems Engineering Research Center (PSerc). The objective was to create a book that contains some cutting-edge research results obtained by the authors, their graduate students and research staff though PSerc and other projects funded by parties outside PSerc. The focus of the book is an overview of new applications in the most critical power system applications during disturbances: state estimation, voltage stability, oscillation monitoring, transient stability, and fault location. Since all the results obtained in this book relate to the use of field data for testing and evaluation, the proposed applications have direct practical value. As such, the book may be used for teaching a graduate course or it may be used by professionals to better understand the opportunities that come with extensive use of time-synchronized measurements.

The introductory chapter of the book provides a history of time-synchronized measurements. It gives a survey of pioneering developments and points out early designs of phasor measurement units (PMUs). The discussions go on to describe other elements of a synchrophasor measurement system: phasor data concentrator and a system for dissemination of the accurate time through the communication network or directly via interfacing PMUs to the Global Positioning System (GPS) of satellite receivers. The motivation for the book is discussed at the end of the chapter.

Chapter 2 is focused on one of the most fundamental power system data acquisition and monitoring concepts, namely state estimation (SE). The history of the Energy Management Systems is outlined and the SE approaches are surveyed. Then, several new concepts in SE are analyzed: substation state estimation, PMU-based state estimation, and quasi-dynamic distributed state estimator. The discussion supports the use of synchronized measurements in daily operations of the power grid.

Chapter 3 deals with the monitoring methods to detect occurrences of instability. The power system stability phenomena covering voltage, transient and small signal stability are surveyed. Various real-time detection methods for monitoring voltage instability, oscillations, and angle instability are demonstrated using archived data from actual power system networks.

Chapter 4 is entirely devoted to the issues of assessment of the transient stability under various power system operating conditions. The use of decision trees to facilitate the transient stability assessment in real-time is emphasized. Some test results from applying such methods in an actual power system contingency are presented. This chapter also gives details of how such an assessment scheme may be implemented.

Chapter 5 is devoted to determining fault location using different time-synchronized concepts, namely the use of synchronized samples and the use of synchrophasors. The initial discussion describes how and why the two techniques may be used to locate faults. For each of the two approaches, both theoretical details and practical aspects are elaborated. The results are also verified using data from an actual power system.

The book opens quite a few research questions and hopefully creates an incentive for the readers to engage in future research that is informed and motivated by the approaches and results described in the book. It also illustrates how such techniques may be implemented and demonstrates the performance using field data. An extensive list of references about prior work is provided at the end of each chapter. The reader can explore various techniques in more detail by reading referenced papers and student dissertations reporting further results from the research teams led by the authors.

We hope this book will be a starting point for many research and implementation directions to be undertaken by the readers in the future.

Authors
Mladen Kezunovic, Texas A&M University
Sakis Meliopoulos, Georgia Institute of Technology
Vaithianathan "Mani" Venkatasubramanian, Washington State University
Vijay Vittal, Arizona State University

Acknowledgments

The authors wish to recognize many graduate students, research staff, and colleagues that contributed to the research results reported in this book. Thanks are due to the following students, research associates, and faculty at Texas A&M University for their contributions included in this book: Dr. Papiya Dutta, Ahad Esmaeilian, Dr. Zijad Galijasevic, Dr. Ashok Gopalakrishnan, Maja Knezev, Prof. Yuan Liao, Dr. Shanshan Luo, Prof. Brana Perunicic, Prof. Alex Sprintson, and Dr. Ce Zheng. The contributions of the following students and research associates at Arizona State University are sincerely appreciated: Siddharth Likhate, Ruisheng Diao, and Kai Sun. Research reported in this book from Washington State University is from the thesis works of graduate students, Guoping Liu, Qiang Zhang, Zaid Tashman, Xing Liu, Michael Sherwood, and Dongchen Hu, and their contributions are gratefully acknowledged. The contributions of the following research associates and students at the Georgia Institute of Technology are acknowledged and appreciated: Prof. George Cokkinides, Dr. Fan Zhang, Dr. George Stefopoulos, Renke Huang and Dr. Evangelos Farantatos.

This book originated from joint work by the authors in a US Department of Energy (USDOE) grant "Future Grid Initiative" under Power System Engineering Research Center (PSERC). The authors thank USDOE and PSERC for facilitating this effort.

Contents

Chapter 1
Introduction

1.1 Background on Synchrophasor Technology

The use of synchronized measurements, particularly synchrophasors, has a history of over 30 years of research and development. This technology allows measurements at different physical locations to be synchronized and time-aligned, then combined to provide a precise, comprehensive view of an entire region or interconnection. Figure 1.1 depicts the locations of currently deployed devices called phasor measurement units (PMUs) that are participating in the North American Synchrophasor Initiative (NASPI) project and provide synchrophasor measurements across North America. It should be noted that there are many more installed PMUs across the USA and around the world that are not part of the NASPI project. In the last few years, the effort of deploying and demonstrating a variety of applications that can benefit from synchronized measurements has been accelerated through the NASPI and other related industry efforts. Most recently, several utilities and regional market operators have developed plans for large-scale deployment of such a technology. In the deployment of the intelligent electronic devices (IEDs) for substation synchronized measurement applications, the focus at present is on two approaches: (a) use of PMUs (dedicated high-precision recording instruments) and (b) use of PMU-enabled IEDs (digital fault recorders (DFRs), digital protective relays (DPRs), digital disturbance recorders (DDRs), and other devices that have PMU measurement capability). While the number of PMUs across the US utility networks in the NASPI network is estimated at 250, the number of PMU-enabled IEDs sold by manufacturers is more than a million and is increasing. With the recent investments through the American Recovery and Reinvestment Act (ARRA) and other funding sources, the total number of PMUs and PMU-enabled IEDs may increase by an order of magnitude as the industry starts utilizing the capabilities of this technology to improve protection, control, and operation of the system in the next 5–10 years. The effective utilization of this valuable asset will require substantial manpower for substation installation, communications, data integration, and visualization.

M. Kezunovic et al., *Application of Time-Synchronized Measurements in Power System Transmission Networks,* Power Electronics and Power Systems,
DOI 10.1007/978-3-319-06218-1_1, © Springer International Publishing Switzerland 2014

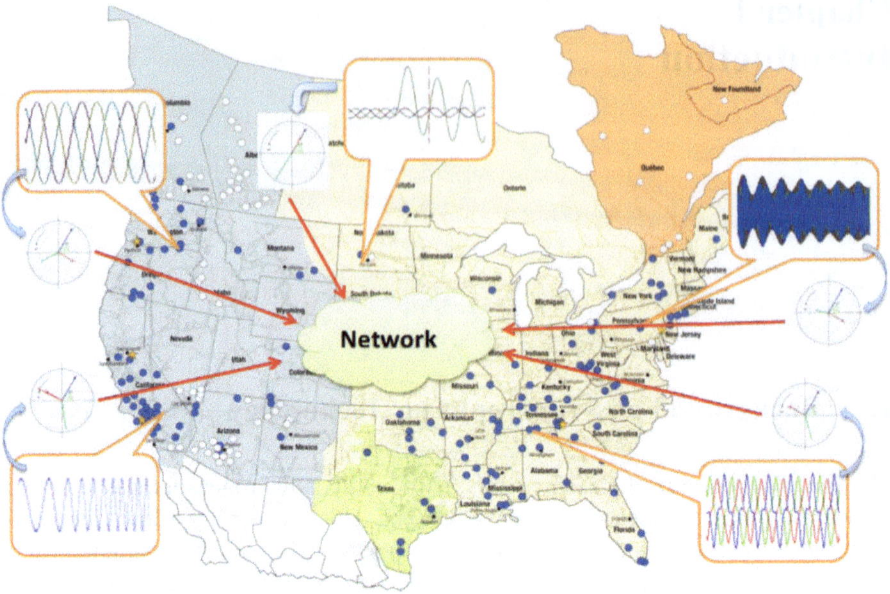

Fig. 1.1 Synchrophasor measurements aggregated across North America

1.2 Synchronized Measurement Devices

The value of synchronized measurements has been recognized in many applications. As a matter of fact, synchronization of measurements and observations was sought in ancient times. Key to this capability is the development of accurate clocks. In ancient times the "klepsidra" (water clock) was typically used for timing. Recently, it has been discovered that the ancient Greeks had also developed a clock mechanism (the antikythera mechanism) that is illustrated in Fig. 1.2. More information about the mechanism can be found in http://www.antikythera-mechanism.gr/. This mechanism consists of a number of geared wheels as in a typical mechanical watch of the twentieth century.

As technology evolved, the accuracy of clocks increased. Today, the most accurate clock is the Cesium atomic clock, with an accuracy of 1 μs in 100 years. The cost of the Cesium atomic clock is prohibitive for many applications. However, the US federal government has developed and deployed a constellation of satellites known as the global positioning system (GPS). Specifically, after years of development and experimental systems, in 1989 the launching of 24 satellites was initiated at about six per year and made available for civilian use. The final GPS satellite was launched on June 26, 1993. The GPS satellites transmit the signal of the Cesium clock throughout the globe and if the signal is received from more than three satellites, using triangulation/estimation methods, the signal of the Cesium atomic clock can be reproduced at any point on earth with an accuracy of tens of nanoseconds.

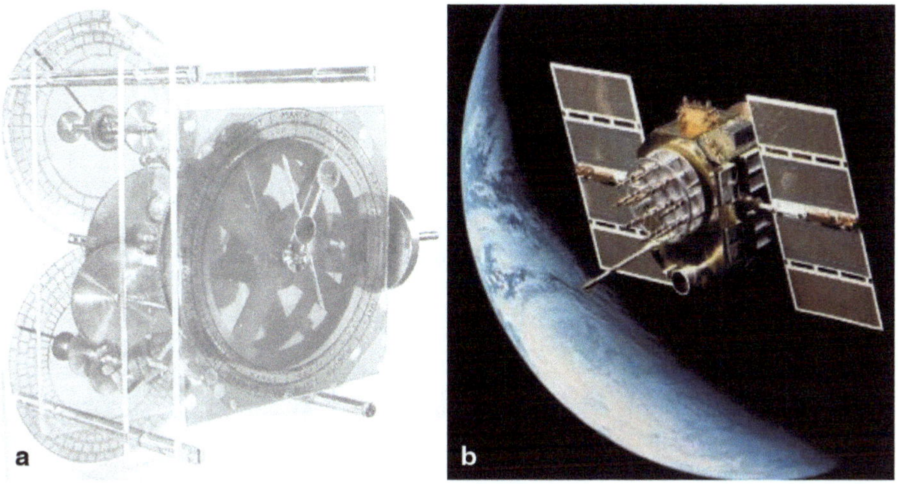

Fig. 1.2 Evolution of time pieces—from the antikythera mechanism (89 AD) (**a**) to the Global Positioning System (1989 AD) (**b**)

Note that from the power system applications point of view a timing signal with accuracy better than 1 μs is sufficient. With proper metering technology, an accuracy of 1 μs translates into measuring the phase angle with an accuracy of 0.02°, an accuracy that is adequate for present day power system applications.

With the introduction of computer relaying (G. Rockefeller, "Fault Protection with a Digital Relay", 1969) [1] and later microprocessor relays (the first commercial microprocessor relay was introduced by Ed Schweitzer in 1983), efforts were initiated to extract the phasors of an electric power system using the available time signals at that time. In 1980, Missout and Girard described a system that was developed for measuring the phase angle between Montreal and Sept-Iles [2]. In 1981, Bonanomi described a system for measuring phase angles using synchronized clocks [3]. In 1983, Meliopoulos [4] proposed to Utah Power company the development of a system of synchronized measurements using timing signals from Geostationary Operational Environmental Satellites (GOES) for monitoring the power loop flow throughout the state of Utah as well as the swings of this power flow in response to concern by this utility for the frequent loop flow problems that they were experiencing. The GOES signal has an accuracy of better than 100 μs and at that time it was the best available clock (other than a Cesium atomic clock). While this proposal was not funded, the authors proposed to New York Power Authority (NYPA) in 1989 to assess and develop the technology of synchronized measurements for the purpose of monitoring harmonic distortion on the NYPA transmission system. This work was funded, and in 1991, a prototype was developed (Meliopoulos/Cokkinides, Georgia Tech Invention Disclosure, 1991) that used signals from a GPS clock and a time-Vernier method to time tag measurements with an accuracy better than 2 μs. The device can be added to an existing non-GPS-synchronized DFR as shown in Fig. 1.3. The principle of operation is also shown in Fig. 1.3.

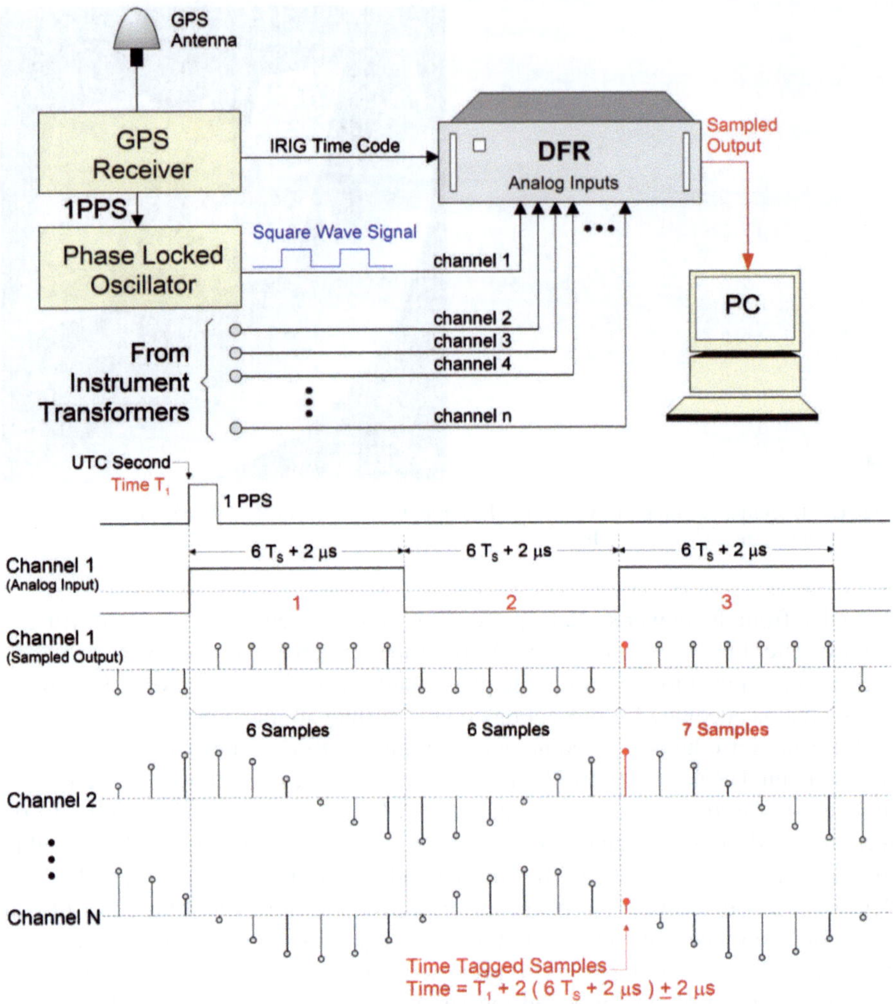

Fig. 1.3 Sample time tagging via time-Vernier scheme (1991). *GPS* global positioning system

A sequence of square pulse signals of a duration of six sampling periods plus 2 μs is generated, which is synchronized with the 1 pps signal of a GPS clock. This signal is then fed into one channel of a DFR. When the number of samples within one square pulse is 7, this indicates that the first and last sample in that pulse is taken at the time of the square pulse transition with an accuracy of 2 μs or better. Subsequently, one of these samples is time tagged as indicated in Fig. 1.3. The samples at the same time of all channels in the DFR are also assigned the same time tag, as seen in the figure. The time tag is then used to compute the Fourier transform of the sampled waveform with reference the GPS signal thus providing the phasor with reference to the GPS clock. This device can be inserted in the input channel

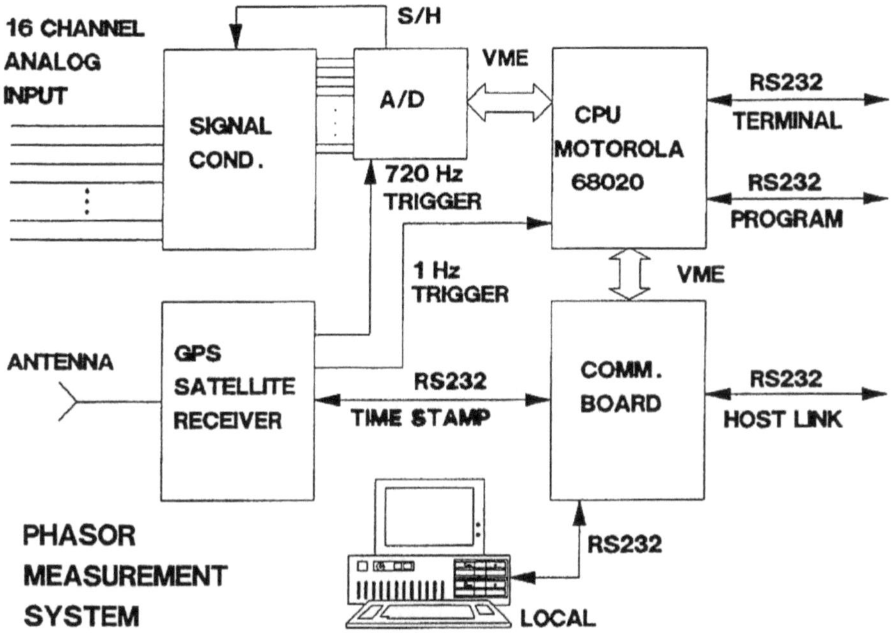

Fig. 1.4 Block diagram of Arun Phadke's PMS (from Phadke [5]). Characteristics: **a** analog anti-aliasing input filter with a cutoff frequency of 360 Hz, **b** 12 bit sample and hold A/D technology, and **c** 720 samples per second with analog multiplexing

of commercially available DFRs, and it is able to time tag the measurements with accuracy better than 2 μs.

During the period 1988–1991, Arun Phadke developed a phasor measurement system (PMS), which is illustrated in Fig. 1.4. The PMS used a GPS signal for timing, 720 samples per second, a sample and hold A/D converter, and a front-end antialiasing filter with a cutoff frequency of 360 Hz. The combination of the anti-aliasing filter and the multiplexing introduced time delays that were orders of magnitude greater than the accuracy of the GPS clock. Although this device was never tested by independent organizations, the estimated timing errors were more than 50 μs. Several PMSs were constructed and sold to utilities (for example, three to AEP, one to NYPA [5]). Despite the use of the GPS clock, the PMS was not capable of performing measurements with comparable accuracy to the GPS clock.

The first device capable of performing synchronized measurements with accuracy comparable to the GPS clock accuracy was developed by Jay Murphy of Macrodyne and was released into the market in January 1992. The device was called the Macrodyne PMU Model 1690. Murphy introduced a number of innovations to achieve the goal of performing synchronized measurements with accuracy comparable to the GPS clock shown in Fig. 1.5. The characteristics of the Macrodyne PMU are: individual channel GPS synchronization (no multiplexing), common mode rejection filter with optical isolation, and a 16 bit A/D sigma/delta modula-

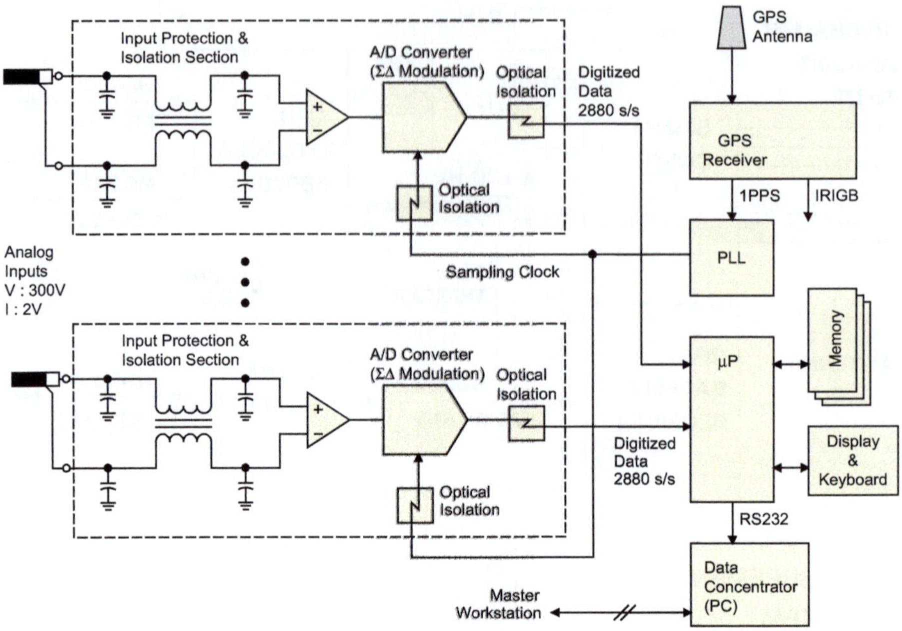

Fig. 1.5 Block diagram of Macrodyne's phasor measurement unit model 1690 (PMU). *PLL* Phase Lock Loop. (Figure created by Meliopoulos based on Model 1690 PMU design document, physical inspection of the internal construction of the unit, and validated by Jay Murphy, 1992)

tion converter operating at several megahertz and decimated to 2880 samples per second.

1.2.1 Standalone PMUs

The need to achieve synchronized measurements for state estimation (SE) has been long recognized. To achieve synchronized measurements, a very accurate clock signal with accuracy better than 1 μs is required at every key measurement location of the power system. Synchronization of the measurements can be achieved with the use of a GPS receiver and appropriate hardware. A conceptual view of such a system is illustrated in Fig. 1.6. The GPS receiver has the capability to provide a synchronization signal with precision better than 1 μs. This time reference allows the measurement of the phase angle of the fundamental frequency component with an accuracy of 0.02° on a system-wide common reference. The local system remote terminal unit (RTU) uses input signals from existing instrument transformers. The captured voltage and current waveforms are time-tagged and transmitted to the energy management system or the master station. Normally, only the first sample needs to be time tagged. Knowing the sampling rate, all other information can be easily extracted. Note that in the energy management system, one can

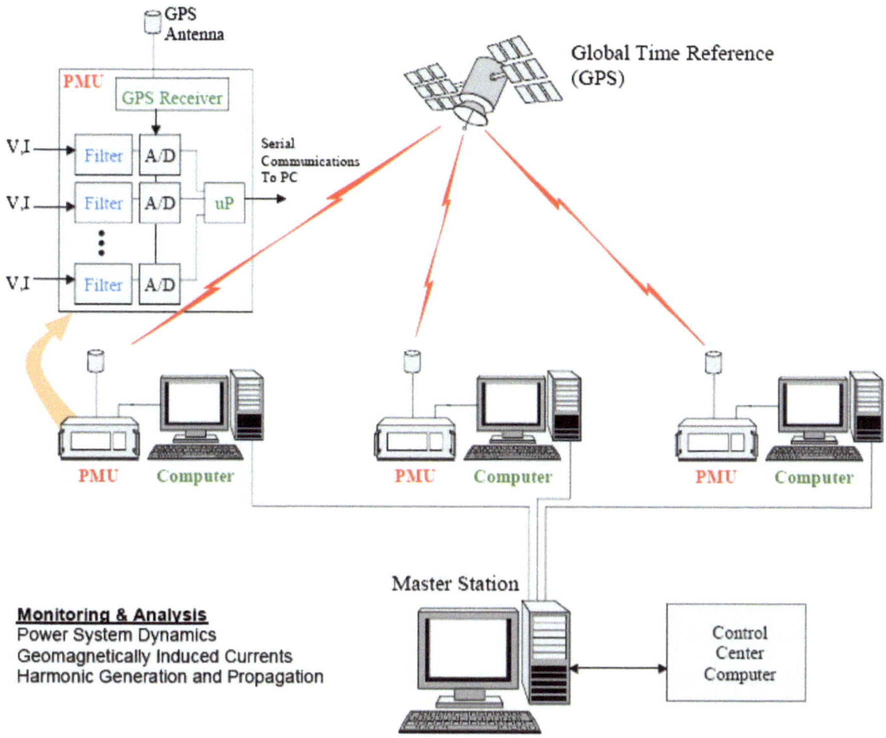

Fig. 1.6 The hardware platform used by the global positioning system (*GPS*)-synchronized measurements. *PMU* phasor measurement unit

collect all the data with the same time tag. The local systems can be programmed to obtain a set of measurements at user-selected intervals with timing accuracy of the GPS signal. A specific standard (IEEE C37.118) [6] exists that defines the standard time intervals of 10, 12, 15, 20, 30, and 60 times per second for 60 Hz systems and 10, 25, and 50 times per second for 50 Hz systems. A time reference provided by GPS can provide a very accurate time reference of better than 1 μs anywhere on earth. Specifically, the phase of voltage and current can be calculated on an almost absolute basis using a highly accurate GPS clock. This time reference allows the measurement of the phase angle of the fundamental with an accuracy of 0.02° on a system-wide common reference.

Presently, commercial PMUs have been installed throughout the eastern, western, and Texas power system interconnections in the USA. Use of synchronized measurements simplifies the SE problem. In most of the cases, PMUs are deployed with several other PMU units to back up other PMUs that might be out of order and deteriorate the computation of SE. Generally, these PMU units are integrated with a number of other functions to build a fully networked and automated power system.

PMUs and PMU-enabled IEDs are typically installed in a substation or at a power plant. Each phasor requires three separate electrical connections (one for each

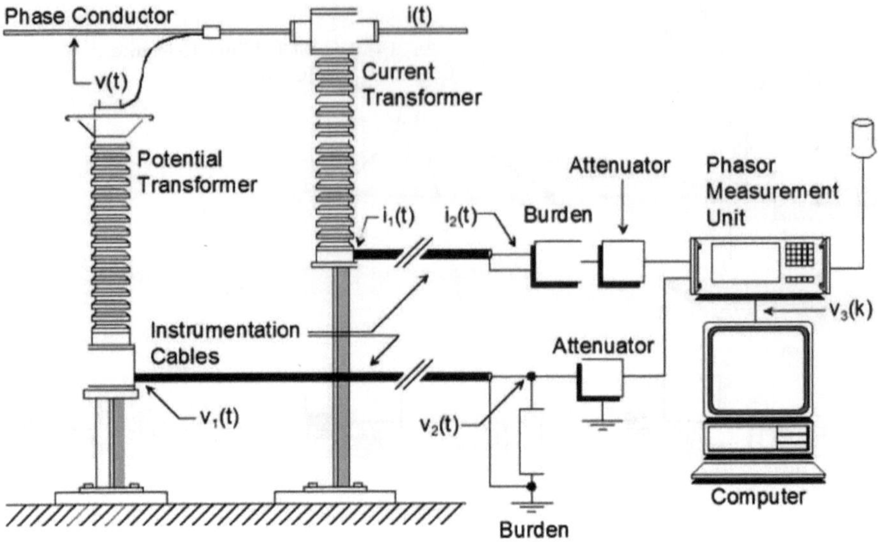

Fig. 1.7 A typical PMU installation—the instrumentation of only one phase is shown in [7]

phase) to either measure a current (from a line or power transformer bank) or a voltage (from either line or bus potential transformers, PTs). A typical PMU installation is shown in Fig. 1.7.

1.2.2 PMU-Enabled IEDs

The synchrophasor measurement functionality need not be the sole function or purpose of a device; for instance, many digital relays have PMU functionality but their primary purpose is to serve as a relay rather than as a PMU. Any device that incorporates this functionality—such as DFRs and digital relays—is considered a PMU device, i.e., a PMU-enabled IED. Other unrelated functions of the device must be shown not to affect the performance of the PMU component, and equally important, the PMU functions must not affect the other functions of the device. The main components of a PMU or PMU-enabled IED include analog input signal interface, data acquisition system, phasor estimation module, and post-processing module for output data. Each module, particularly the phasor estimation algorithm has an impact on performance accuracy. For the use of such devices in various applications, the synchrophasor and frequency values must meet the general definition as well as the minimum accuracy requirements given in standards [6, 8, 9].

Presently many IEDs have PMU functionality. PMUs output real-time streaming synchrophasor data usually in synchrophasor format (see IEEE Std C37.118 [6]) and at various rates, such as 10, 12, 15, 20, and 30 for 60 Hz systems. PMUs

Fig. 1.8 PMU-enabled IEDs from different manufacturers

may also provide positive, negative, zero sequence values, frequency, and rate of change of frequency (ROCOF). Recently many IEDs have been available on the market with PMU functionality. Phasor measurement is an added function to the primary functions of a device. Most of the relay manufacturers are adding PMU functionality in their line of protective relays, meters, and fault recorders. As various manufacturers use different algorithms for the phasor calculation, the resulting synchrophasor data may have variations in accuracy and latency. A picture of several PMU-enabled IEDs from different manufacturers is depicted in Fig. 1.8.

As of today, there are over a million DPRs and DFRs deployed on the US grid with PMU functionality. These PMU-enabled IEDs can be used for high-speed grid monitoring, automated operations, forensic analysis, and model calibration. By the end of 2013, NIST reported that there were at least 50 commercially available IEEE C37.118 compatible, PMU-capable, upgradable, relays, DPRs, and DFRs offered by 20 manufacturers [10].

1.2.3 Time Synchronization Options

A PMU requires a source of universal time coordinated UTC time and high accuracy timing signals to provide synchronized measurements. According to the IEEE C37.118 standard, the accuracy of a synchrophasor measurement is defined in terms of the total vector error (TVE, see Fig. 1.9), which shall not exceed 1 % of TVE.

If half of the TVE comes from magnitude measurement error and the other half comes from phase measurement error, then the phase angle error is 0.28°. If only

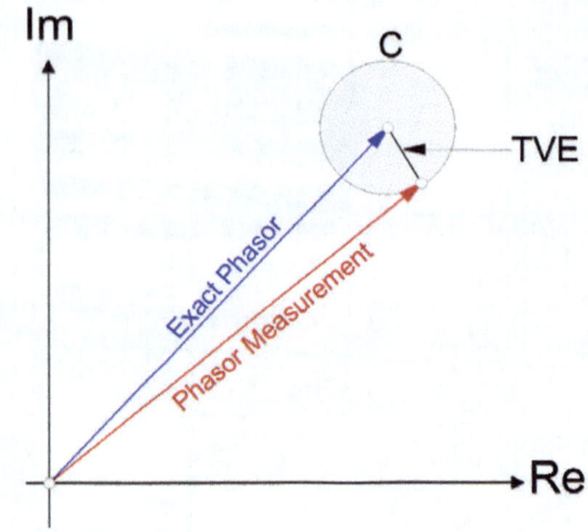

Fig. 1.9 Definition of total vector error (*TVE*)

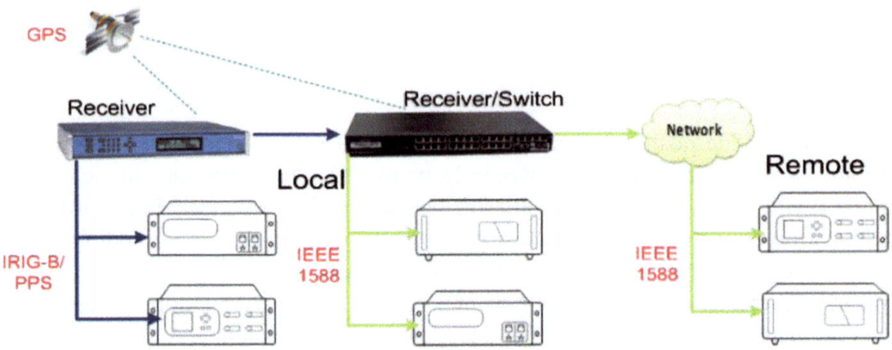

Fig. 1.10 Time synchronization options. *GPS* global positioning system

the phase angle error is considered, then the error of 0.28° corresponds to approximately 13 μs error at 60 Hz and 16 μs error at 50 Hz. The existing synchronization methods include the direct GPS signal [11], IRIG-B/PPS [11], and IEEE 1588 [12]. For using the direct GPS signal, an IED must be equipped with a GPS receiver for decoding the time signal. For using IRIG-B and PPS, the receiver must be local to the IEDs. Using IEEE PC37.238 [13], the receiver can be either local or remote to the IEDs because the time code defined in IEEE PC37.238 can be distributed over a communication network as shown in Fig. 1.10.

GPS signal—A direct way for a PMU to reference UTC time is by using a GPS signal. Such PMUs are specially designed to receive and synchronize a local timing reference to UTC using the GPS signal. The GPS system is referenced to atomic clocks maintained by the US military. These clocks are very accurate time and frequency references, and are maintained within a known offset to UTC. A good GPS time receiver will incorporate averaging and holdover to accommodate short deviations and signal dropouts due to atmospheric disturbances and other causes. It will also provide notification if it loses lock with the GPS signal, so the synchronization status is always known to PMUs .

IRIG-B/PPS—IRIG-B is commonly used by PMUs for synchronizing to UTC time. It may be provided in a level shift, a 1 kHz amplitude modulated signal, or in the bi-phase Manchester modulated format (modulation type 2, B2xx). If the amplitude modulation is used, it may need to be supplemented with a 1 PPS pulse train to achieve the required accuracy. The IRIG-B amplitude modulated format is commonly available and hence is the most readily implemented. The newer Manchester format is more compatible with fiber optic and digital systems and provides complete synchronization without additional signals. The clock reference is provided once per second in seconds through day of year in a binary coded decimal (BCD) format and an optional binary second-of-day count. A PPS in timing systems is a pulse train of positive pulses at a rate of 1 Hz. The rising edge of the pulses coincides with the seconds change in the clock and provides a very precise time reference. The pulse widths vary from 5 μs to 0.5 s, and the signal is usually a 5 V amplitude driving a 50 ohm load.

IEEE 1588 PTP—is a protocol used to synchronize distributed clocks with an accuracy of less than 1 μs via Ethernet networks. GPS is a highly accurate solution but does not scale well due to cost and complications of attaching antennas to every device. Using an Ethernet network to propagate timing signals eliminates the extra cabling requirement of GPS and IRIG-B. In the first version of IEEE 1588, it was found that the original design does not scale well for large switched networks. The processing and traffic congestion can occur at the timing master, and cascaded network switches could cause inaccuracies due to jitter that occurs as the packets traverse through the switches. The second generation of this protocol solves these problems by (a) using mechanisms that increase accuracy by accounting for switching time and peer-to-peer propagation delays that occur as the timing signals traverse the network and (b) using "transparent clocks" in Ethernet switches that eliminate the need for end-to-end delay measurement, reducing traffic congestion and eliminating switch jitter. With IEEE 1588, the cabling infrastructure requirement is reduced. The convergence of timing and data information networks can be carried out right to the network edge and converted to IRIG-B for synchronization of existing devices that are not capable of IEEE 1588, allowing them to be kept in service even while updating the timing and data network infrastructure.

1.2.4 Phasor Data Concentrators

A phasor data concentrator (PDC) collects phasor data from multiple PMUs or other PDCs, aligns the data using time-tag to create a time-synchronized dataset, and passes this dataset to other information systems. A PDC also performs data quality checks and missing flags or problematic data (waiting for a set period of time, if needed, for all the data to come in before sending the aggregated dataset on). Some

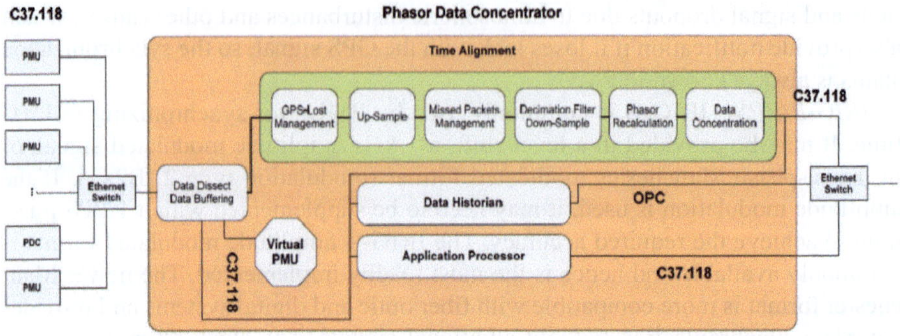

Fig. 1.11 Phasor data concentrator. *GPS* global positioning system, *PMU* phasor measurement unit, *PDC* phasor data concentrator, *OPC* OLE for process control

PDCs also store and down-sample (decimate) phasor data points so that phasor data can be fed directly to applications that use data at slower sample rates, such as a supervisory control and data acquisition (SCADA) system. The real time data broadcasting capability of PDCs enables other utilities to make use of the time aligned data. Each measurement carries a precise time stamp taken from GPS satellites, so that the entire electric grid can be analyzed at any moment in time. PDC supports the synchrophasor standard, C37.118, (or the older synchrophasor standard, IEEE 1344), PDC stream, and object linking and embedding (OLE) for process control (OPC) for real time data transmission. OPC and extensible markup language (XML) formatted historical data access options enable us to store and access massive amounts of data. An example of a PDC with internal functions is illustrated in Fig. 1.11.

The IEEE PES Power System Relaying Committee (PSRC) has developed two standards that further clarify PMU measurement requirements and, define the functional and performance requirements for PDCs [14, 15]. The main functions include the ability to

- correlate phasor data by time-tag and then broadcast the combined data to other systems
- conform to streaming protocol standards (e.g., IEEE C37.118.1) for both the phasor data inputs and the combined data output stream
- verify the integrity and completeness of data streams from PMUs and properly handle data anomalies
- buffer input data streams to accommodate the differing times of data delivery from each PMU

The functions of a PDC can vary depending on its role or its location between the source PMUs and the higher-level applications. There are three levels of PDCs, as shown in Fig. 1.12.

- *Local PDC*—manages the collection and communication of time-synchronized data from local PMUs, sends it to higher-level concentrators, and stores the data for use within the substation. A local PDC may be located physically close to

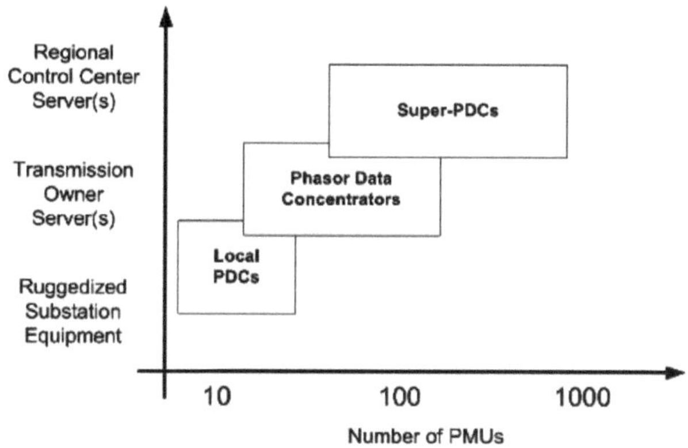

Fig. 1.12 Three levels of phasor data concentrators (*PDCs*) [13]. *PMU* phasor measurement unit

PMUs (typically at a substation) and store a small cache of local measurements to prevent against network failure; it should also be the source of data for local automated control functions. A local PDC is generally a hardware-based device that typically requires limited maintenance and can operate independently if it loses communication with the rest of the synchrophasor network .

- *Corporate PDC*—operates within a control room where it aggregates data from multiple PMUs and PDCs. It must conduct real-time data quality checks and calculations at very high speed with real-time sample rates currently at 10 to 60 samples per second and heading higher. Real time calculation must be completed very quickly before the next set of values arrives. PDCs typically utilize threading and other parallel computing techniques available within modern operating systems to manage multiple connections at high speeds. PDCs must be adaptable to new protocols and output formats as well as interfaces with data-using applications. PDCs should incorporate inter-utility standards to allow for efficient machine-to-machine, program-to-program compatibility. It is expected that grid operators who host PDCs will demand these capabilities from PDC vendors, and that these capabilities will be written into specifications and tested for effective interoperability performance before acceptance .
- *Super PDC*—operates on a regional scale, handling phasor measurements from several hundred PMUs and multiple PDCs. It collects and correlates phasor data from remote PDCs and PMUs and makes them available as a coherent, time-synchronized dataset to applications such as wide-area monitoring and visualization software, energy management systems, and supervisory control and data acquisition SCADA applications. A super-PDC also feeds the data into a central database for long-term data archiving (data historian function). Super-PDCs are software implementations, running on mainstream server hardware, as these larger devices need to scale rapidly to serve growing utility and regional deployment of PMUs, as well as diverse phasor data applications .

PDCs are commercially available from several vendors. Based on their performance history, they are generally considered to be production-grade systems. Even

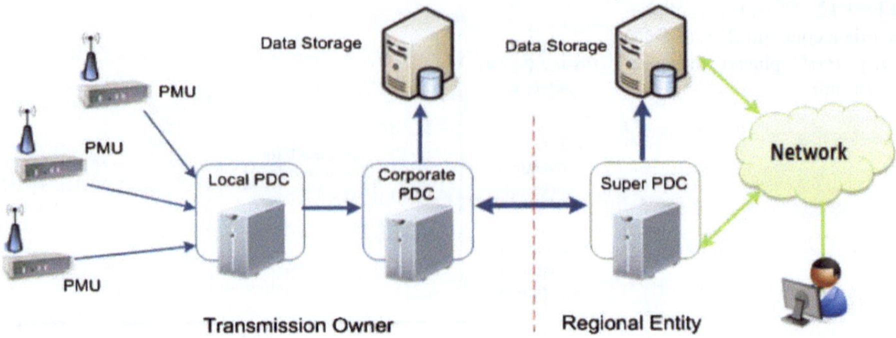

Fig. 1.13 A basic synchrophasor network. *PMU* phasor measurement unit, *PDC* phasor data concentrator

so, PDCs have not yet been performance-benchmarked. While it is likely that the smart grid interoperability group SGIG projects will accelerate the development of PDC performance and test standards, there are currently no formal standards for evaluating and rating PDCs.

1.3 Synchronized Measurement Networks

A basic synchrophasor measurement network consists of PMUs and PMU-enabled IEDs, PDCs, gateways, data storage, and applications, as given in Fig. 1.13. The IEEE PES Power System Relaying Committee (PSRC) has adopted a standard for synchrophasor data transfer for power systems [15]. It defines the format of synchrophasor data and specifies the requirements of network for data transfer. The functions of PMU and PDC have been described in previous sections.

1.4 Motivation for the Book

From the brief introduction provided in the preceding sections of this chapter, it is clear that synchronized measurements are being adopted and deployed at an accelerated pace in power systems around the world. One of the commonly mentioned applications of synchronized measurements is improved situational awareness for power system operators. Given the rich information available in the measurements, improved situational awareness is not a significant application to justify return on investment made in obtaining the measurements. Hence, this textbook develops various applications of synchronized measurements for a range of operating goals to enhance power system reliability and efficiency. Motivated by such goals, the authors have addressed several issues that will help the reader to better understand and apply techniques that use synchronized samples and/or synchronized phasors.

Chapter 1 gives a brief history of the concept of time-synchronized measurements for uses in power system applications. The basic elements of a system solution are described including the PMUs, the PMU-enabled IEDs, the time synchronization sources, and PDCs. The role of each of the solution elements is elaborated and how they all fit together is illustrated. This chapter enables the reader to grasp the synchrophasor technology and its intended uses.

Chapter 2 details the use of synchronized phasor measurements to enhance power system SE. This is a critical function in real time power system operation and drives many other critical applications. The evolution of the state estimator function from the legacy state estimators developed in the 1960s to the present fully distributed state estimators (substation based state estimator, QSE) is presented. The impact and role of GPS-synchronized measurements on state estimators are discussed.

An overview of the different concepts of instability mechanisms, namely, voltage instability, oscillatory instability, and angle instability, are introduced in Chap. 3. Real-time algorithms for detecting these three stability phenomena are then discussed in Chap. 3 along with examples from simulated test data as well as archived PMU data from real power systems.

In many utilities, operations' planning is based on a close to real time model of the system, which is regularly updated to track the changes in system topology and change operating conditions. In such situations synchronized phasor measurements can be effectively used in conjunction with the operations planning model to determine operating limits for rotor angle stability, voltage stability, and small-signal stability. The approach presented in Chap. 4 is based on conducting off-line simulations close to real time to determine operating limits based on the available synchronized phasor measurements. A decision tree is then trained using the phasor measurements as critical attributes. The real time phasor measurements are then used to determine when the system is approaching. Two large test cases are presented as examples. In the first case, the proposed method is applied to the Entergy test system to determine transient and small-signal stability. In the second test case, the proposed method is applied to determine voltage stability .

Chapter 5 presents and details the use of synchronized measurements for fault location. For this application, a differentiation between uses of synchronized samples versus synchronized phasors is made since the two algorithms discussed use different inputs. The use of synchronized samples is based on an ability to detect and classify faults correctly, and on knowledge of the differential equation model of the line, and as such, these two issues are explored in detail. The use of synchronized phasors is focused on the technique that utilizes sparse measurements due to an ability to have access to synchronized phasors measured at two ends of the line. This technique is sensitive to fault resistance assumptions. Hence, the two issues, namely the fundamentals of the proposed algorithm and the sensitivity to the fault resistance are discussed in detail. Each of the techniques covered in this chapter is illustrated with some realistic test cases based on field measurements and simulations.

References

1. G. D. Rockefeller, Fault protection with a digital computer. IEEE Trans. Power Appar. Syst. **PAS-88**(4), 438–464 (1969)
2. G. Missout, P. Girard, Measurement of bus voltage angle between Montreal and Sept-Iles. IEEE Trans. Power Appar. Syst. **PAS-99**(2), 536–539 (1980)
3. P. Bonanomi, Phase angle measurements with synchronized clocks. IEEE Trans. Power Appar. Syst. **PAS-100**(12), 5036–5043 (1981)
4. A. P. Meliopoulos, Phase angle difference monitoring via GOES synchronized measurements. Proposal to EPRI, 1983
5. A. G. Phadke, Synchronized phasor measurements—a historical overview. Presentation at the transmission and distribution conference and exhibition 2002, Asia Pacific, vol. 1, pp. 476–479, 6–10 Oct, 2002
6. IEEE Standard C37.118–2005, IEEE Standard for synchrophasors for power systems, 2005
7. A. P. Sakis Meliopoulos, G.J. Cokkinides, F. Galvan, B. Fardanesh, P. Myrda, Delivering accurate and timely data to all: model based substation automation applications for advanced data availability. IEEE Power Energy Mag. **5**(3), 74–86 (2007)
8. IEEE C37.242–2013, IEEE guide for synchronization, calibration, testing, and installation of phasor measurement units (PMUs) for power system protection and control, 2013
9. IEEE C37.244–2013, IEEE guide for phasor data concentrator requirements for power system protection, control and monitoring, 2013
10. IEEE Tutorial, Implementation of synchrophasor systems. Organizer: Vahid Madani. IEEE PES conference on innovative smart grid technologies, 19–22 Feb, Washington. D.C., 2014
11. IRIG Standard 200−04, IRIG serial time code formats, Range Commanders Council, US Army White Sands Missile Range, New Mexico, Sept 2004
12. IEEE Std. 1588–2008, IEEE standard for a precision clock synchronization protocol for networked measurement and control systems, 2008
13. IEEE Std. C37.238–2011, IEEE standard profile for use of IEEE 1588 precision time protocol in power system applications, 2011
14. IEEE C37.118.1–2011, IEEE Standard for Synchrophasor Measurements, 2011
15. IEEE C37.118.2–2011, IEEE Standard for Synchrophasors Data Transfer for Power Systems, 2011

Chapter 2
State Estimation and Visualization

One obvious application of GPS-synchronized measurements is the dynamic monitoring of the operating conditions of the system or the dynamic state estimation of the system. Since GPS-synchronized measurements can directly measure magnitude and phase of an electrical quantity (voltage and current) and since the state of the system often is defined as the voltage phasors at all buses of the system, some researchers have declared that GPS-synchronized measurements provide for the direct measurement of the system state and therefore there is no need for state estimation. This is not quite right since GPS-synchronized measurements are always tainted with measurement errors, and procedures for filtering the errors are necessary. These procedures are enabled by state estimators. Another important reason for using state estimators is that they provide the mathematical framework for validating the measurements against the model of the components of the system. This is the only practical approach to validate measurements. GPS-synchronized measurements enable new, better, accurate, and faster procedures for state estimation. In other words, GPS-synchronized measurements are the enabling technology to achieve system visibility and awareness at speeds not imagined before.

Most control and operation functions in a control center are model based. A model of the system is used to determine the best control action, or the model is used to determine the effects of specific actions (using a "what if analysis"). Since state estimation utilizing GPS-synchronized measurements provides a more accurate and reliable model of the system at faster speeds than before, more accurate applications and potentially dramatic improvements in the reliability and effectiveness of the controls are enabled.

In this chapter, the overall energy management infrastructure of a power system is reviewed, the role of the state estimator and the impact of GPS-synchronized measurements on the state estimator are discussed.

M. Kezunovic et al., *Application of Time-Synchronized Measurements in Power System Transmission Networks,* Power Electronics and Power Systems,
DOI 10.1007/978-3-319-06218-1_2,

2.1 The Energy Management System

The electric power system is a complex geographically dispersed system that delivers energy to most vital parts of the modern infrastructure of residential, commercial, and industrial facilities. It plays a very important role in the economic activities of any country and is responsible for generating, transmitting, and delivering energy to all vital economic activities of the society. The electric infrastructure system is evolving as more and more economic activities are "electrified" such as the recent trend of electrifying the transportation sector via electric vehicles and plug-in hybrid vehicles. Over time, the complexity of the system has grown. To manage this complex system, monitoring, control, and operation functions are computer assisted. Computer control systems of electric power systems have evolved as computer and monitoring technology evolved. Throughout the years, these systems have been named "control centers," "energy management systems (EMSs)," and "independent system operations." The names reflect the changing emphasis in the functions of these control centers. For simplicity, the term EMS will be used.

The EMS concept comprises hardware and software for the purpose of monitoring and controlling the power system. Typically, the function of monitoring is fully automated and comprises many sensors and meters that are typically integrated into a digital system. The introduction of GPS-synchronized measurements created devices that take measurements synchronized in time with microsecond precision. These measurements are very valuable to the overall monitoring system of a power system. As will be seen later, they provide new and better ways to identify the system model in real time at much faster computational speeds than previous technologies. In this chapter, an examination of how this technology has changed the process of state estimation and the process of extraction of the real time model of the system will be presented. The monitoring function and the subsequent extraction of the real time model of the system is fundamental to the control of the power system because the control methodology of power systems is model based. In general, control functions are either automated or manual, but they are both (automatic or manual) model based, i.e., the model is utilized to determine the best control option and to determine the effect of the control action on the system before the control is exercised.

Historically, EMSs have evolved from the offices of traditional dispatchers, who have in their reach supervisory equipment. Based on their experience, they would monitor the supervisory equipment and control the system appropriately. The control was manually executed upon communication between the dispatcher and local operator (e.g., plant operator). As the size and complexity of the system grew, this approach was not adequate. A number of incidents indicated that the security of the system, defined as the ability to operate in synchronism under possible random disturbances, could not be guaranteed with this simple approach. Out of this need, a comprehensive and integrated approach to monitor and control a power system has emerged.

Advances in computers and power system hardware have opened up new possibilities. As the technology has evolved, so has the EMS. A modern EMS is characterized by:

1. Monitoring system—which evolved from the basic supervisory control and data acquisition (SCADA) system to a digital monitoring system with state estimation, bad data detection, identification and rejection, validation of the real time model of the system, and sophisticated visualization methods of the system operating conditions.
2. Dispatch operation—which has been replaced with the fully digital automatic generation control (AGC). The AGC integrates the dispatch function with the load frequency control, power interchange control problem, and various power system optimization functions.
3. System security functions (monitoring and control)—which have been integrated into a hierarchical control scheme.
4. Advanced economy scheduling functions—which are an integral part of the system, including access to the power markets.

The hardware required for the new approach is illustrated in Fig. 2.1. The sensors and controls, located in the field, collect data. For example, a sensor can be a wattmeter, a voltmeter, a current meter, or a breaker status device. These data are collected at the remote terminal units (RTUs), which are normally hardwired to the sensors and controls. Then, the data are transferred through communication channels to communication input/output controllers (CIOCs) located normally in a central location. There the data are transferred to the computers. Computer programs evaluate the data, use them for state estimation, detect and reject bad data, provide the real time model of the system, and display the information on computer screens, dynamic mimic boards, or computer-generated projections of system displays using sophisticated visualization tools. The operator can visualize the operation of the system by examining the displays.

In a modern EMS, a computer (automatically or on dispatcher command) can issue commands, which are then transferred through the CIOCs, communication links, and RTUs to the field equipment for execution (e.g., a trip command for a breaker).

The hardware configuration of Fig. 2.1 provides the possibility of controlling and operating the system in a rather sophisticated manner. The number of functions and controls is relatively large. Table 2.1 provides a partial list of the computational and control functions. As mentioned, these control functions are model based. The state estimation provides the real time model that is needed by these functions. This is the reason state estimation stands out from the other functions. In this chapter, the focus is on state estimation and in particular the utilization of phasor measurement unit (PMU) data in the state estimation function.

2.2 Real Time Operational Requirements

Effective control and operation of electric power systems requires accurate and reliable knowledge of the system model and the operating state of the system in real time. For this purpose, modern power systems are equipped with an extensive data

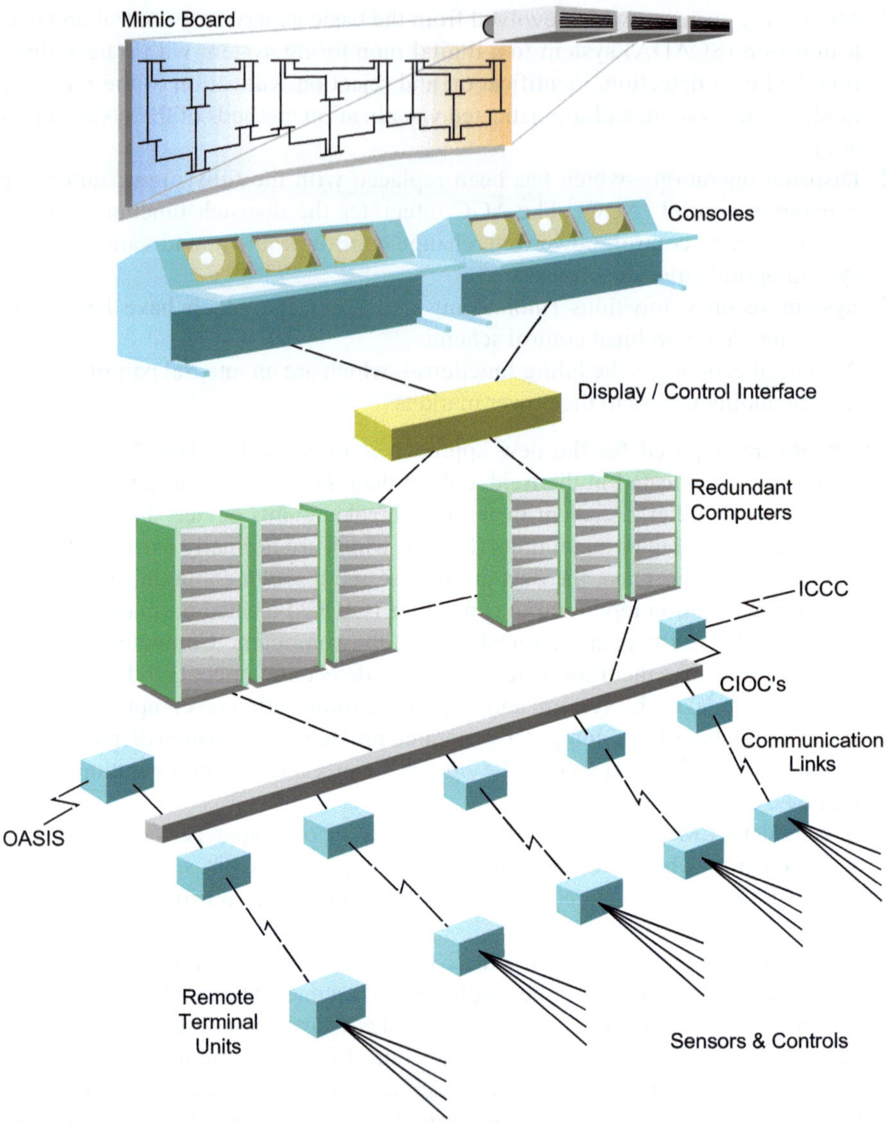

Fig. 2.1 Typical configuration of energy management system hardware

acquisition system. Local analog and status quantities, such as voltage magnitude, real and reactive power flows, loads, and status of breakers (open/close) are measured and transmitted to a central location. The measurements are simple, requiring simple instrumentation. Typical analog measurements are: (a) voltage magnitudes, (b) real and reactive power flows, and (c) current magnitude measurements. Recent technology based on GPS has made it possible to measure voltage and current phase

Table 2.1 Partial list of computational and control functions

Data acquisition and processing	Energy management (AGC)	Security monitoring and control
SCADA	Automatic generation control	Contingency selection
Network topology	Economic dispatch	Contingency analysis
State estimation	Optimal power flow	Voltage stability assessment
Visualization	Unit commitment	Dynamic stability assessment
	Load forecast	Security controls
	Wind/PV forecast	
	Interchange evaluation	
	Locational marginal prices	

PV photovoltaic, *SCADA* supervisory control and data acquisition, *AGC* automatic generation control

angles as well. Typical status measurements are: (a) breaker status and (b) disconnect switch status. The traditional measurements are taken every one to several seconds. The PMU measurements are recorded several times per second and most typically 30 or 60 times per second as discussed in the previous chapter. Measurements are transmitted to a central location (the EMS or the energy control center) where they are processed to yield the operating state of the system. The process consists of two analysis problems: (a) determination of network topology and (b) determination of operating state. The network topology is constructed from the status of breakers and disconnect switches. The operating state of the system is constructed from analog measurements by means of state estimation. The state estimator addresses two issues. First, measurements are usually corrupted with error (resulting from potential transformer or current transformer inaccuracies, instrument error, transmission error, calibration, and human error in setting instrument parameters.). These errors are directly transmitted to the computed operating state. State estimation enables the filtering of the measurement errors and the extraction of the best estimate of the system model. Second, it is possible that one of the measurements may include a large error (gross error—due to meter malfunction or communication errors) resulting in a large discrepancy between the computed operating state and the actual operating state of the system. The state estimator enables the detection and identification of wrong measurements or errors in measurements (bad data).

The state estimator provides an ideal approach for the solution of the above-mentioned problems by taking advantage of redundant measurements. The redundant measurements are utilized to compute the best estimate of the operating state of the system based on statistical methods. As discussed in subsequent sections of this chapter, state estimation provides the mechanism to: (a) determine whether the system state can be computed from existing data (observability); (b) filter out usual measurement errors and, therefore, compute the system state with minimum error; (c) identify and reject bad data; and (d) determine the degree of confidence on the estimated state of the system. The conceptual view of the process is illustrated in Fig. 2.2. The figure illustrates two sets of input data to the state estimator: (a) status data and (b) analog data. The status data are used by the topology processor to construct the system topology. The analog data are subjected to consistency checks, limit checks, and detection of data with large errors. The filtered data and

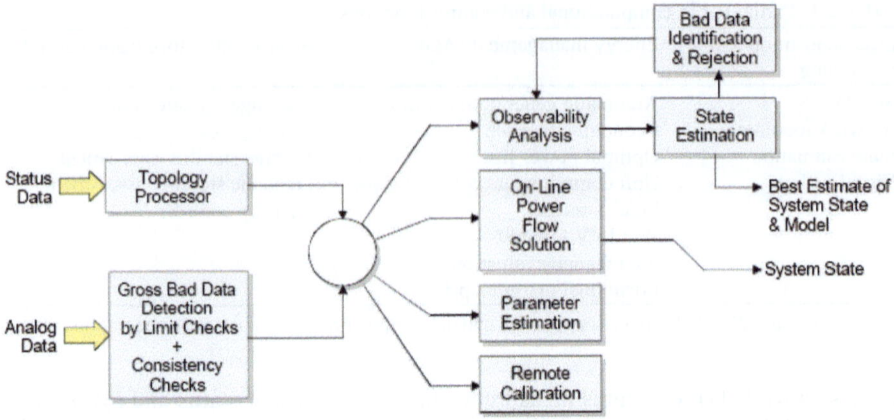

Fig. 2.2 Conceptual view of real time power system modeling and state estimation

the topology are used in the observability analysis to determine whether the state of the system can be computed from this data. Subsequently the state estimation is performed followed by bad data detection and identification. The end result is a validated real time model of the system.

In addition, other applications are possible using the collected data. These applications are parameter estimation and remote calibration of the meters as indicated in Fig. 2.2. The figure also indicates the on-line power flow option. The on-line power flow uses a subset of the available data to define and solve a power flow problem. The constituent parts of the state estimation process will be discussed next.

2.3 SCADA System

The necessary hardware to enable real time modeling of a power system is collectively referred to as the SCADA system. The supervisory control subsystem consists of hardware and software, which (a) collect status data (e.g., breaker status open/close) and analog data (e.g., measurements of voltage magnitude, current, real power, and reactive power) and transmit these data to a central location for processing and display and (b) allow remote tripping of breakers or changes of transformer tap. In most cases, supervisory control is a manual function, e.g., the dispatcher at the control center will initiate a command to open/close a breaker. The data-acquisition subsystem consists of remote terminal equipment for interfacing with power system instrumentation and control devices; interfaces with communication channels; and equipment for interfacing with the system control center. Since the SCADA system transmits data from the field to a central location and vice versa, communication media, protocols, and communication speeds are very important. In the past, it was customary to have separate communication channels for the SCADA. Today, however, it is a unified system, sharing common two-way communication channels that may consist of several physical layers. Communications are integrated into the RTU, which manages data collection, control functions, and communication with a master station. The master station (which is typically located in the control center) has multiple communication channels to RTUs. Many times a dedicated channel is assigned to each RTU. In other cases, there are fewer channels than RTUs, requiring more than one RTU to share a channel. Analog data is scanned periodically, typically every one second to a few seconds. Each scan is triggered by the system control center at the prescribed interval by using a request to all remote stations to send in data. The amount of data collected and transmitted is very large for typical power systems. These data are transported via communication channels. In order to minimize communication traffic some form of data compression is utilized. For example, for status data one can send only changes of status data. This approach minimizes the amount of data transfer and the amount of processing needed at the master station. Analog data can be also compressed with a number of methods. Independent of system configuration, SCADA system manufacturer, communication software, and computer configuration, the end result of the SCADA system function will be a collection of a set of system data for every sampling period. The data typically consists of:

- Breaker status
- Disconnect switch status
- Transformer tap setting
- MW flow measurements
- MVAr flow measurements
- Voltage magnitude (kV) measurements
- Current magnitude (kA) measurements
- Phase angle difference measurements

A simplified view of a SCADA system is illustrated in Figs. 2.3 and 2.4.

At the central location (EMS or control center), the data are managed using the data acquisition software, which performs the following tasks:

1. Data collection initiation and placement in the computer memory
2. Gross error checking
3. Conversion to engineering units
4. Limit checking
5. Generation of a data base interfaced with application programs

With the introduction of numerical relays and GPS-synchronized measurements (PMUs) the architecture of the RTU has changed. For example, the RTU may be a single intelligent electronic device (IED) connected to a "station bus," which is connected to a number of devices, e.g., relays and PMUs, or it can be a PDC that collects data from a number of PMUs and transmits this data to the master station. In fact, recently, merging units were introduced that allow for an additional flexibility in the configuration. Figure 2.5 illustrates some of these possibilities. In a typical

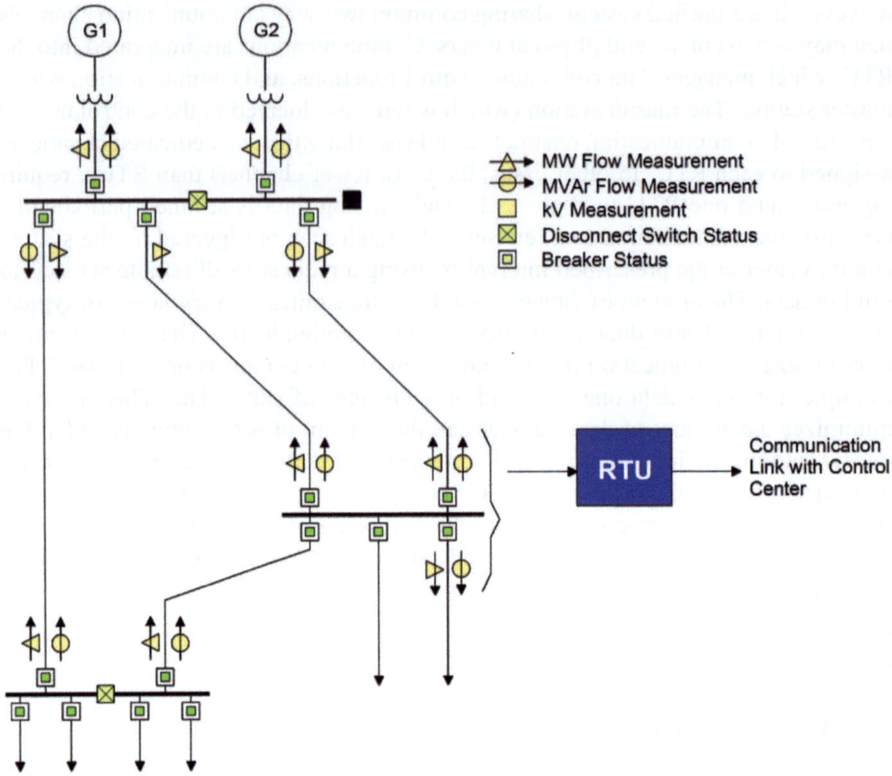

Fig. 2.3 Simplified view of a SCADA system—survey points

Fig. 2.4 Simplified view of a SCADA system—configuration

substation, one may observe subsystems of different eras integrated into the overall SCADA system. This reality is partially illustrated in Fig. 2.5.

The communication protocols used for PMU data and traditional SCADA data are different. Specifically, PMU data are streaming data in a format defined in the synchrophasor standard C37.118 [1]. The traditional SCADA data may use a number of legacy protocols or any of the new standards such as IEC 61850. In any case, the mixing of the data requires that the communications architecture should support

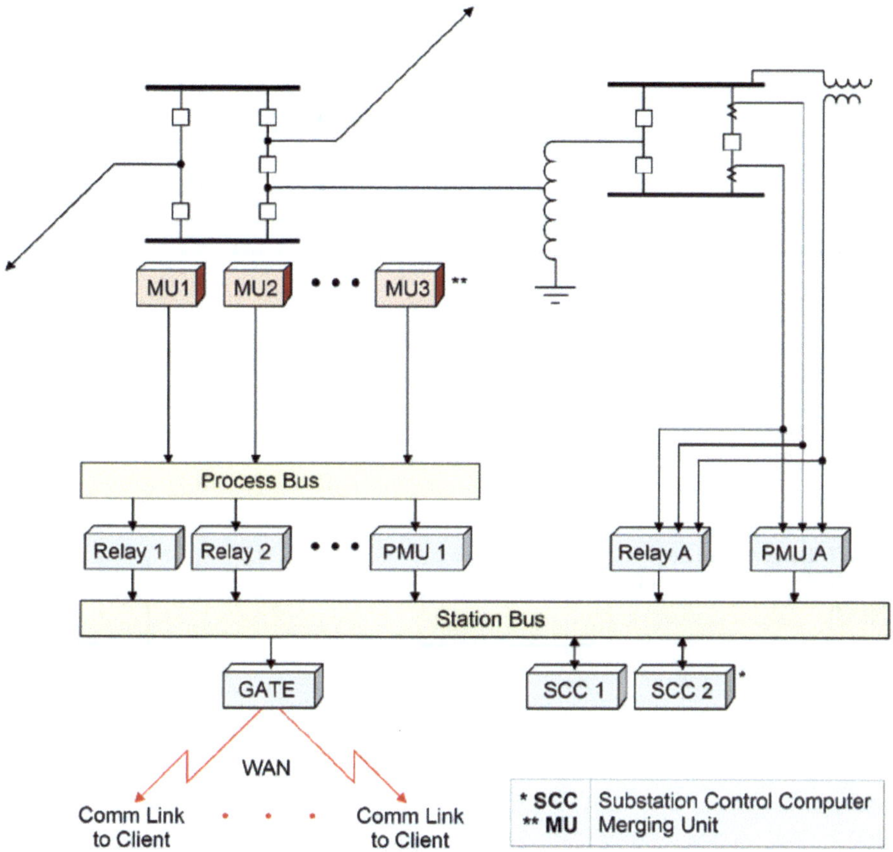

Fig. 2.5 Data collection, controls, and communications at a substation (Merging units/process bus/station bus architecture and numerical relays/station bus architecture)

the various standards or the entire substation should be designed in an integrated manner. This topic is beyond the scope of this book.

The data are utilized to form the system model (network configurator) and to estimate the system operating state (state estimation). The following sections describe these applications.

2.4 System Network Configurator

Data collected with the SCADA system are utilized in two ways. Status data (e.g., circuit breaker status, interrupt switch status, and transformer tap setting) are utilized to form the system network configuration and model. The software that takes

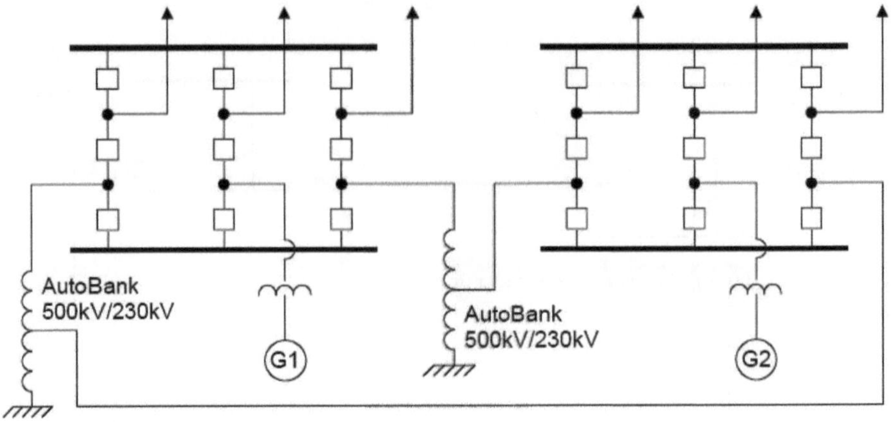

Fig. 2.6 Breaker-oriented model—pre-stored network data

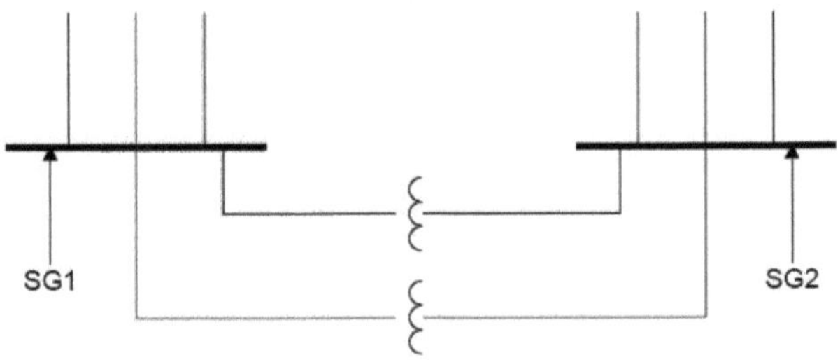

Fig. 2.7 Bus-oriented network model of the system of Fig. 2.6

the status data and computes the system network configuration and model is known as the system network configurator. The information received with the SCADA system determines the status of the breakers. The system network configurator uses pre-stored information (physical arrangement of breakers, switches, and other substation equipment) and the breaker/switch status to determine a "bus oriented" model, i.e., which circuits are connected to which bus and what is the model of each circuit. A simple example is illustrated in Figs. 2.6 and 2.7. Figure 2.6 illustrates the breaker-oriented model at a substation. This model is invariant over time and it can be referred to as pre-stored network data. It changes only when new additions are implemented, such as addition of new breakers, construction of new lines, and other additions. Assuming all breakers are in the close position, Fig. 2.7 illustrates the bus-oriented model of the system. Typically, this procedure is executed only when a change in status data occurs.

The system network configuration and model is next combined with the analog data for the purpose of determining the operating conditions of the system. Typically, there are redundant measurements, which are used to obtain the best (in some

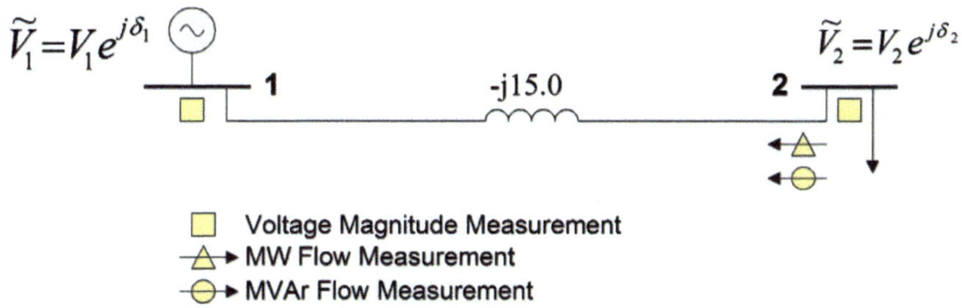

Fig. 2.8 A simplified example for power system state estimation

sense to be discussed later) estimate of the operating state of the system. The computational procedure that performs this task is known as the state estimator and will be discussed next.

2.5 Legacy State Estimation Algorithms

State estimation is a computational procedure, which uses a redundant set of measurements and a bus-oriented network model to compute a statistical estimate of the system operating state. In a power system, using the legacy state estimation formulation, the operating conditions are uniquely defined by the set of variables consisting of all bus voltage magnitudes and angles except the phase angle at an arbitrarily selected bus, which is set equal to zero. We refer to this set of variables as the state of the system. As an example, the operating condition of the system of Fig. 2.8 is defined as having three state variables: V_1, V_2, and δ_2. Knowledge of these three variables is sufficient to determine other quantities of interest, e.g., P_{12} and Q_{12}. For this example, assume that a set of redundant measurements is taken as it is shown in Fig. 2.8. (four measurements: V_1, V_2, P_{21}, and Q_{21}). A subset of these measurements is enough to provide the state variables. For example, assume that the following three measurements are selected: V_1, P_{21}, and V_2, where P_{21} is the real power flow on circuit $2-1$. From these three measurements, the state variables V_1, V_2, and δ_2 can be computed by simply formulating the power flow problem for this system and then solving the power flow problem to obtain the phase angle at bus 2.

In general, a measurable quantity such as P_{21}, Q_{21}, P_{12}, Q_{12}, and V_1, can be expressed as a function of the system state. Let z_i denote a measured quantity. Then:

$$z_i = h_i(x), \tag{2.1}$$

where x is the system state and h_i is a function specific to the measured quantity z_i. Assume that m measurements are taken. Then the entire set of measurements can be expressed as the following vector equation:

$$z = h(x), \tag{2.2}$$

where
x is the system state—an $n \times 1$ vector,
z is a vector of measured quantities—an $m \times 1$ vector, and
h is a vector function—an $m \times 1$ vector function.

Typically, more measurements are taken than the number of state variables to be determined, i.e., $m > n$. In this case, Eq. (2.2) represents an over-determined set of nonlinear equations in real variables. In general, an over-determined set of equations, such as the one described does not have a solution for x. Only if the system model expressed with the functions $h(x)$ is "exact" and the measurements z are extremely "accurate," then Eq. (2.2) has a unique solution for x. This, of course, is unlikely in a real system. Yet, it is possible to obtain a solution for x via a procedure known as the state estimation. The formulation of the problem is illustrated with an example.

Example E2.1 Consider the simplified power system of Fig. 2.8. The following measurements are taken (all in p.u.):

$$z_1 = V_1 = 1.0, \qquad z_2 = V_2 = 0.99$$
$$z_3 = P_{21} = -1.56, \qquad z_4 = Q_{21} = 0.01.$$

Formulate the state estimation problem as an over-determined set of nonlinear equations.

Solution The state vector x is:

$$x = \begin{bmatrix} V_1 \\ V_2 \\ \delta_2 \end{bmatrix}.$$

The vector function $h(x)$ is given as:

$$\begin{aligned}
h_1(x) &= V_1 \\
h_2(x) &= V_2 \\
h_3(x) &= 15.0 V_1 V_2 \sin\delta_2 \\
h_4(x) &= 15.0 V_2^2 - 15.0 V_1 V_2 \cos\delta_2.
\end{aligned}$$

The state estimation problem is formulated as the following over-determined set of equations:

$$
\begin{aligned}
1.0 &= V_1 \\
0.99 &= V_2 \\
-1.56 &= 15.0V_1V_2 \sin\delta_2 \\
0.01 &= 15.0V_2^2 - 15.0V_1V_2 \cos\delta_2.
\end{aligned}
$$

In summary, the state estimation problem is formulated as an over-determined set of nonlinear equations. Specifically, the problem is posed as follows:

Given

$$b = h(x),$$

where

b	is a known $m \times 1$ vector,
$h(x)$	is an m vector of known function,
x	is the unknown $n \times 1$ vector, and
$m > n$.	

Compute the vector x.

This problem is known as the nonlinear estimation problem. In the case that the functions h are linear, then the problem collapses to the linear estimation problem defined as follows:

Given

$$b = Hx,$$

where

b	is a known $m \times 1$ vector,
H	is a known $m \times n$ matrix,
x	is the unknown $n \times 1$ vector, and
$m > n$.	

Compute the vector x.

In general, the over-determined Eq. (2.2) does not have a solution, i.e., a vector x does not exist that exactly satisfies Eq. (2.2). In this case, it is expedient to define the residual vector:

$$r = h(x) - b \text{ (nonlinear case)}$$

or

$$r = Hx - b \text{ (linear case)}$$

and compute a solution x that will minimize the residual vector in some sense. The residual vector represents the error of the measurement, i.e., the error introduced by the meter by which the measurement was taken. The measurement error many times is represented with the Greek letter η. Thus, the above equations can be also written as:

$$r = \eta = h(x) - b \text{ (nonlinear case)}$$

or

$$r = \eta = Hx - b \text{ (linear case).}$$

For the solution of this problem, three approaches have been extensively researched in the power industry: (a) the weighted least squares solution, (b) the weighted least absolute deviation solution (L_1 approach), and (c) the Chebyshev or min–max solution (L_∞ approach). In this book, we will focus on the first approach, which is the most widely used and is introduced next.

The weighted least squares solution of the over-determined system above is the vector x, which minimizes the sum of the weighted squares of the components of the residual vector r or the vector of the measurement errors η. Mathematically, this is expressed as follows:

$$\text{Minimize } J = \sum_{i=1}^{m} w_i r_i^2 = r^T W r = \eta^T W \eta,$$

where

w_i the weight for the residual r_i,
W a diagonal matrix, the diagonal elements being the weights w_i.

Note that the weights are selected so that they will weigh more accurate measurements over less accurate measurements. The most logical selection is to select the weights as the reciprocal of the variance of the measurement error of the meter, i.e.,

$$w_i = 1/\sigma_i^2,$$

σ_i is the standard deviation of the meter expected error.

Note that with the above selection the weighted least square problem can be cast in terms of the normalized measurement error, defined as:

$$s_i = \frac{\eta_i}{\sigma_i}.$$

In this case, the problem is stated as follows:

$$\text{Minimize} \quad J = \sum_{i=1}^{m} \left(\frac{h_i(x) - z_i}{\sigma_i} \right)^2 = \sum_{i=1}^{m} s_i^2 = \eta^T W \eta,$$

where

$$W = \text{diag}\left\{ \frac{1}{\sigma_1^2}, \frac{1}{\sigma_2^2}, \ldots, \frac{1}{\sigma_m^2} \right\}.$$

Note that this is a weighted least squares formulation with the weights defined as the inverse of the squared standard deviations. It should be also noted that a simple least squares approach can be followed by simply assuming all weights to be 1.0. This implies that all standard deviations of the error of all measurements are equal.

Using the nonlinear and linear model equations, the nonlinear and linear state estimation problem is expressed as follows:

$$\text{Minimize} \quad J = \left[h(x) - b \right]^T W \left[h(x) - b \right], \text{ nonlinear case,} \tag{2.3}$$

$$\text{Minimize} \quad J = (Hx - b)^T W (Hx - b), \text{ linear case.} \tag{2.4}$$

The unknown vector x is obtained from the solution of the necessary conditions, which in matrix notation are expressed as follows:

$$\frac{dJ}{dx} = 0.$$

The above problem is first solved for the linear case and then for the nonlinear case.

Linear Case Direct differentiation of Eq. (2.4) with respect to x, we obtain:

$$\frac{dJ}{dx} = \frac{d}{dx}\left[(Hx - b)^T W (Hx - b)\right] = 2H^T W (Hx - b) = 0.$$

The solution of the last equation for the state vector x yields

$$x = \left(H^T W H\right)^{-1} H^T W b.$$

The above equation provides the solution to the linear estimation problem.

Nonlinear Case To obtain the solution to the nonlinear estimation problem, assume that an initial guess of the vector x^0 is known. The nonlinear model equations are linearized around the point x^0 yielding:

$$r = \eta = h\left(x^0\right) + \frac{\partial h(x)}{\partial x}\Big|_{x=x^0} \left(x - x^0\right) + h.o.t. - b,$$

where $h.o.t.$ denotes higher order terms. Assuming that the vector x^0 is very close to the solution, then the $h.o.t.$ are negligibly small and are omitted from above equation, yielding:

$$r = \eta = \frac{\partial h(x)}{\partial x}\Big|_{x=x^0} \left(x - x^0\right) + h\left(x^0\right) - b.$$

Let

$$\frac{\partial h(x)}{\partial x}\Big|_{x=x^0} = H, \text{ and } b' = -h\left(x^0\right)| + Hx^0 + b.$$

Observe that the vector $b' = -h(x^0) + H x^0 + b$ is known (or computable). Now obtain:

$$r = \eta = Hx - b'.$$

Now the problem is identical to the linear estimation problem. Thus, the solution is:

$$x = \left(H^T W H\right)^{-1} H^T W b'.$$

Upon substitution of the b vector

$$x = \left(H^T W H\right)^{-1} H^T W \left[H x^0 - h\left(x^0\right) + b \right] = x^0 - \left(H^T W H\right)^{-1} H^T W \left[h\left(x^0\right) - b \right]$$

The last equation is generalized into the following iterative equation:

$$x^{v+1} = x^v - \left(H^T W H\right)^{-1} H^T W \left[h\left(x^v\right) - b \right],$$

where H is the matrix $\partial h(x) \partial x$ computed at $x = x^v$. This is the Jacobian of the vector function $h(x)$.

In summary, the weighted least squares solution of the linear estimation problem is a direct solution (no iterations) and the weighted least squares solution of the nonlinear estimation problem can be obtained with the iterative algorithm above. An example will illustrate the method.

Example E2.2 Consider the over-determined set of equations derived in Example E2.1. Compute the solution $x^T = [V_1 \ \ V_2 \ \ \delta_2]$, using as an initial guess $x^{0^T} = [1.0 \ \ 0.99 \ \ 0.0]$ and least squares estimation. Assume that all weights are equal to 1.0.

Solution The problem will be solved with the iterative algorithm. Since the weights are all 1.0, the weight matrix W is the identity matrix. The objective function is:

$$J = \left(V_1 - 1.0\right)^2 + \left(V_2 - 0.99\right)^2 + \left(15 V_1 V_2 \sin\delta_2 + 1.56\right)^2 + \left(15 V_2^2 - 15 V_1 V_2 \cos\delta_2 - 0.01\right)^2$$

$$\text{Min } J = \sum_{i=1}^{4} w_i \left[h_i(x) - z_i \right]^2.$$

The Jacobian matrix H is:

$$H = \begin{bmatrix} 1 & 0 & 0 \\ 0 & 1 & 0 \\ 15 V_2 \sin\delta_2 & 15 V_1 \sin\delta_2 & 15 V_1 V_2 \cos\delta_2 \\ -15 V_2 \cos\delta_2 & 30 V_2 - 15 V_1 \cos\delta_2 & 15 V_1 V_2 \sin\delta \end{bmatrix}.$$

The computations follow:

1st iteration

$$r^0 = h\left(x^0\right) - b = \begin{bmatrix} 0.0 \\ 0.0 \\ 1.56 \\ -0.1585 \end{bmatrix}$$

$$H\left(x^{0}\right)=\begin{bmatrix} 1.0 & 0.0 & 0.0 \\ 0.0 & 1.0 & 0.0 \\ 0.0 & 0.0 & 14.85 \\ -14.85 & 14.7 & 0.0 \end{bmatrix}$$

$$H^{T}H=\begin{bmatrix} 221.5225 & -218.295 & 0.0 \\ -218.295 & 217.09 & 0.0 \\ 0.0 & 0.0 & 220.5225 \end{bmatrix}$$

$$H^{T}r^{0}=\begin{bmatrix} 2.3537 \\ -2.33 \\ 23.166 \end{bmatrix}$$

$$\left(H^{T}H\right)^{-1}H^{T}r^{0}=\begin{bmatrix} 0.00538 \\ -0.00532 \\ 0.10505 \end{bmatrix}$$

$$x^{1}=\begin{bmatrix} 1.0 \\ 0.99 \\ 0.0 \end{bmatrix}-\begin{bmatrix} 0.00538 \\ -0.00532 \\ 0.10505 \end{bmatrix}=\begin{bmatrix} 0.9946 \\ 0.9953 \\ -0.10505 \end{bmatrix},$$

2nd iteration

$$r^{1}=h\left(x^{1}\right)-b=\begin{bmatrix} -0.0054 \\ 0.0053 \\ 0.0030 \\ 0.0823 \end{bmatrix}$$

$$H\left(x^{1}\right)=\begin{bmatrix} 1.0 & 0.0 & 0.0 \\ 0.0 & 1.0 & 0.0 \\ -1.5655 & -1.5644 & 14.767 \\ -14.8472 & 15.0222 & -1.557 \end{bmatrix}$$

$$H^{T}H=\begin{bmatrix} 223.89 & -220.589 & -0.000647 \\ -220.589 & 229.114 & -46.4911 \\ -0.000647 & -46.4911 & 220.489 \end{bmatrix}$$

$$H^T r^1 = \begin{bmatrix} -1.2320 \\ 1.2369 \\ -0.0838 \end{bmatrix}$$

$$\left(H^T H\right)^{-1} = \begin{bmatrix} 0.4960 & 0.4988 & 0.1052 \\ 0.4988 & 0.5063 & 0.1068 \\ 0.1052 & 0.1068 & 0.0270 \end{bmatrix}$$

$$\left(H^T H\right)^{-1} H^T r^1 = \begin{bmatrix} -0.0027869 \\ 0.0027593 \\ 0.00019540 \end{bmatrix}$$

$$x^2 = \begin{bmatrix} 0.99741 \\ 0.99255 \\ -0.10525 \end{bmatrix}.$$

This completes the solution. It is expedient to compute the residuals (or measurement errors) using the above computed state:

$$\begin{aligned}
r_1 &= V_1 - 1.0 = -0.00259 \\
r_2 &= V_2 - 0.99 = -0.00256 \\
r_3 &= 15.0 V_1 V_2 \sin\delta_2 + 1.56 = 0.00000614 \\
r_4 &= 15.0 V_2^2 - 15.0 V_1 V_2 \cos\delta_2 - 0.01 = 0.0000566.
\end{aligned}$$

Note that in this case the residuals or measurement errors are very small.

Statistical Properties of the State Estimator The state estimator has a number of statistical properties that are very useful in quantifying the quality of the state estimate. The statistical properties are derived from the fact that the normalized measurement error used in the weighted least squares approach can be assumed to be Gaussian distributed. Specifically the normalized error (residual),

$$s_i = \frac{h_i(x) - b_i}{\sigma_i},$$

is assumed to be a Gaussian random variable with expected value zero and variance 1.0. Details of the statistical properties can be found in [2] and in numerous books on state estimation. Some additional details are also provided in the section "Substation-based state estimation." Here, we describe two important applications of the statistical properties of the state estimators without proof. These are the chi-square test and the covariance of the estimated states.

The chi-square test provides the probability that the measurements fit the model of the system. This is often referred to as "goodness of fit" or "confidence level." The name chi-square test comes from the fact that the sum of the normalized residuals squared is a chi-square random variable, assuming the normalized residuals are Gaussian random variables. The chi-square test is applied as follows:

Step 1. Compute the state estimate $\hat{x}$ in the weighted least squares sense.

Step 2. Compute the value $\zeta = \sum_{i=1}^{m} s_i^2(\hat{x}) = \sum_{i=1}^{m} \left(\frac{h_i(\hat{x})-b_i}{\sigma_i}\right)^2$.

Step 3. "Read" the probability $\boldsymbol{Pr}\left[\chi^2 \le \zeta\right] = \boldsymbol{Pr}(\zeta, v)$ from the chi-square probability distribution function.

Step 4. Compute the probability $\boldsymbol{Pr}\left[\chi^2 \ge \zeta\right] = 1.0 - \boldsymbol{Pr}(\zeta, v)$.

Above probability is the probability that measurements fit the model of the system. This can be also expressed as a confidence level. The meaning of this probability is that a low value indicates that the set of measurements may contain bad data, i.e., data that do not fit the model. The probability value is typically used to detect bad data in the measurement set, i.e., low value indicates presence of bad data.

The covariance of the state estimates is given by:

$$C_x = \left(H^T W H\right)^{-1}.$$

The expected standard deviation of a state estimate is:

$$\sigma_{x_i} = \sqrt{C_x(i,i)},$$

where $C_x(i,i)$ is the ith diagonal entry of the covariance matrix.

The legacy state estimator algorithm described above uses nonsynchronized data that have some time skews, which can introduce biases in the state estimator. Other sources of biases come from model inaccuracies, e.g., the legacy state estimator is based on a positive sequence model that may have an error up to 5 % from the actual system. Another important source of bias errors is that the legacy state estimator uses measurements from one phase only. Any imbalance in the system will create biased state estimator results. For more information, consult [2]. These observations indicate the need for better state estimators. Over the years state estimators have evolved. GPS-synchronized measurements have significantly improved state estimators. Key developments in this area are:

a. Inclusion of GPS-synchronized measurements into the legacy state estimator was suggested in 1985 [3] and 1986 [4]. Reference [3] describes the legacy state estimator with nonsynchronized and synchronized measurements (nonlinear state estimator), while [4] describes the estimator with only synchronized measurements (linear state estimator; LSE). Both of these estimators have the same biases and errors described above [2].
b. The first three-phase LSE was formulated in Cartesian Coordinates, developed and installed on the NYPA system in 1993–1998, following the introduction of

the GPS system in 1989 and the development of GPS-synchronized measurements [5–7].

c. The three-phase state estimator in (b) was further developed into the quadratized distributed dynamic state estimator or substation-based state estimator [8–13]. These developments will be described next.

2.6 PMU-Based LSE

In Chap. 1, the GPS-synchronized measurement technology was introduced. These measurements are capable of providing the magnitude and phase of voltage and currents with high accuracy and synchronized to a high-precision clock (GPS signal). Use of synchronized measurements simplifies the state estimation problem. As a matter of fact, by using GPS-synchronized measurements and a linear model of the system, the state estimator becomes linear (LSE) [5]. The PMU-based LSE will be introduced next.

Consider, for example, the case in which all measurements are GPS-synchronized. We assume here that the GPS-synchronized measurements have a time precision of better than 1 μs, the capability of the GPS technology. Note that in practice this precision may not be achievable due to other sources of error (e.g., [14]). Consider a set of GPS-synchronized measurements $\tilde{z}$. These measurements are typically in the form of voltage phasor measurements (phase A to neutral, phase B to neutral, and phase C to neutral) and electric current phasor measurements (phases A, B, and C). Let $\tilde{x}$ be the vector of the bus voltage phasors (states, three complex states per bus, phases A, B, and C). The measured voltage and electric current phasors will be functions of the voltage phasors $\tilde{x}$. In general, these functions will be linear. They can be written as:

$$\tilde{z} = Y\tilde{x} + \tilde{\eta},$$

where Y is a constant matrix and $\tilde{\eta}$ is the measurement error.

The state estimation problem is now formulated as follows:

$$\text{Min } J = \sum_i \left|\frac{\tilde{\eta}_i}{\sigma_i}\right|^2 = \tilde{\eta}^H W \tilde{\eta},$$

where $\tilde{\eta} = \tilde{z} - Y\tilde{x}$, and superscript H denotes hermitian (transposition and conjugation).

The above state estimation problem is linear. The solution is obtained by first separating the equations into real and imaginary parts yielding:

$$\text{Min } J = \eta_r^T W \eta_r + \eta_i^T W \eta_i.$$

Subject to:

$$\begin{aligned} \eta_r &= z_r - Y_r x_r + Y_i x_i \\ \eta_i &= z_i - Y_r x_i - Y_i x_r \end{aligned},$$

where the subscript r indicates real and the subscript i indicates imaginary.

Upon substitution and observing that at the solution the derivatives of the function J, with respect to the state variables, will vanish:

$$\frac{dJ}{dx_r} = -2Y_r^T W\left(z_r - Y_r x_r + Y_i x_i\right) - 2Y_i^T W\left(z_i - Y_r x_i + Y_i x_r\right) = 0,$$

$$\frac{dJ}{dx_i} = -2Y_i^T W\left(z_r - Y_r x_r + Y_i x_i\right) - 2Y_r^T W\left(z_i - Y_r x_i + Y_i x_r\right) = 0.$$

Solution of the above equations for the state vector provides:

$$\begin{bmatrix} x_r \\ x_i \end{bmatrix} = \begin{bmatrix} A & B \\ -B & A \end{bmatrix}^{-1} \begin{bmatrix} Y_r^T W z_r + Y_i^T W z_i \\ Y_r^T W z_i - Y_i^T W z_r \end{bmatrix},$$

where

$$\begin{aligned} A &= Y_r^T W Y_r + Y_i^T W Y_i \\ B &= -Y_r^T W Y_i + Y_i^T W Y_r \end{aligned}.$$

The linear state estimation problem with GPS-synchronized measurements will be demonstrated with a simplified example.

Example E2.3 Consider the two-bus, single-transmission-line system of Fig. E2.3. Assume the following synchronized measurements were taken (note they are in the form of phasors):

$$\tilde{V}_1 = 437.621e^{+j61.46^0}\ kV \qquad \tilde{V}_2 = 432.508e^{+j55.09^0}\ kV$$

$$\tilde{I}_1 = 0.6424e^{+j73.95^0}\ kA \qquad \tilde{I}_2 = 0.6601e^{-j143.96^0}\ kA$$

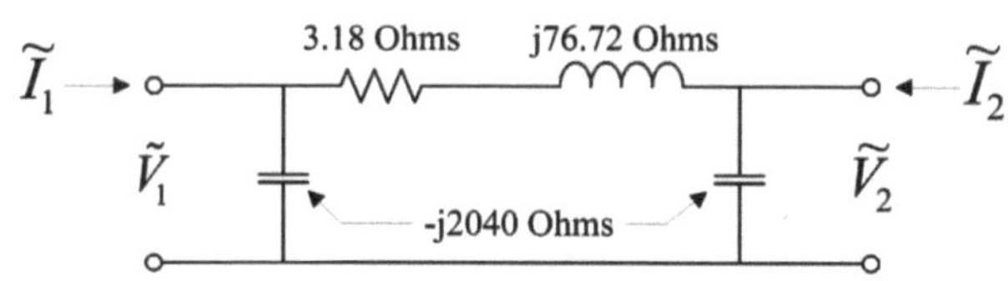

E2.3

The voltage meters have an accuracy of 2 % on a 440 kV scale and the electric current meters have an accuracy of 1 % on a 1500 A scale. Formulate the state estimation problem using Cartesian coordinates, i.e., assume that the state of the system is defined in terms of the real and imaginary part of the bus voltages, i.e.,

$$x = \left[x_1\ x_2\ x_3\ x_4\right]^T,\ \tilde{V}_1 = x_1 + jx_2, \text{ and } \tilde{V}_2 = x_3 + jx_4.$$

a. Compute the best estimate of the state in the weighted least squares sense.
b. Compute the confidence level (probability of model and measurement fit).
c. Compute the standard deviations of the state variables.

Solution The admittances are:

$$g + jb = \frac{1}{3.18 + j76.72} = 0.00054 - j0.01301S,$$

$$jb' = \frac{1}{-j2040.0} = j0.00049S.$$

a. The electric currents as functions of voltages are:

$$\begin{aligned}\tilde{I}_1 &= (0.00054 - j0.01252)\tilde{V}_1 - (0.00054 - j0.01301)\tilde{V}_2 \\ \tilde{I}_2 &= -(0.00054 - j0.01301)\tilde{V}_1 + (0.00054 - j0.01252)\tilde{V}_2\end{aligned}.$$

The measurements as functions of the state variables are:

$$\begin{aligned}
&\boldsymbol{Re}\{\tilde{V}_1\} = 209.08 = x_1 \\
&\boldsymbol{Im}\{\tilde{V}_1\} = 384.44 = x_2 \\
&\boldsymbol{Re}\{\tilde{V}_2\} = 228.62 = x_3 \\
&\boldsymbol{Im}\{\tilde{V}_2\} = 367.15 = x_4 \\
&\boldsymbol{Re}\{\tilde{I}_1\} = 0.1776 = 0.00054x_1 - 0.01252x_2 - 0.00054x_3 + 0.01301x_4 \\
&\boldsymbol{Im}\{\tilde{I}_1\} = 0.6173 = -0.01252x_1 + 0.00054x_2 + 0.01301x_3 - 0.00054x_4 \\
&\boldsymbol{Re}\{\tilde{I}_2\} = -0.4738 = -0.00054x_1 + 0.01301x_2 + 0.00054x_3 - 0.01252x_4 \\
&\boldsymbol{Im}\{\tilde{I}_2\} = -0.3642 = 0.01301x_1 - 0.00054x_2 - 0.01252x_3 + 0.00054x_4
\end{aligned}$$

In compact form: $z = Hx$,
where

$$z = \begin{bmatrix} 209.08 & 384.44 & 247.515 & 354.6724 & 0.1776 & 0.6173 & -0.5338 & -0.3884 \end{bmatrix}^T,$$

$$H = \begin{bmatrix} 1 & 0 & 0 & 0 \\ 0 & 1 & 0 & 0 \\ 0 & 0 & 1 & 0 \\ 0 & 0 & 0 & 1 \\ 0.00054 & 0.01252 & -0.00054 & 0.01301 \\ -0.01252 & 0.00054 & 0.01301 & -0.00054 \\ -0.00054 & -0.01301 & 0.00054 & 0.01252 \\ 0.01301 & -0.00054 & -0.01252 & 0.00054 \end{bmatrix},$$

$$x = \begin{bmatrix} x_1 & x_2 & x_3 & x_4 \end{bmatrix}^T.$$

The state estimation solution is obtained from (direct solution)

$$x = \left(H^T W H\right)^{-1} H^T W z,$$

where

$$W = \begin{bmatrix} 0.0129 & 0 & 0 & 0 & 0 & 0 & 0 & 0 \\ 0 & 0.0129 & 0 & 0 & 0 & 0 & 0 & 0 \\ 0 & 0 & 0.0129 & 0 & 0 & 0 & 0 & 0 \\ 0 & 0 & 0 & 0.0129 & 0 & 0 & 0 & 0 \\ 0 & 0 & 0 & 0 & 4444.44 & 0 & 0 & 0 \\ 0 & 0 & 0 & 0 & 0 & 4444.44 & 0 & 0 \\ 0 & 0 & 0 & 0 & 0 & 0 & 4444.44 & 0 \\ 0 & 0 & 0 & 0 & 0 & 0 & 0 & 4444.44 \end{bmatrix},$$

$$x = \left(H^T W H\right)^{-1} H^T W z = \begin{bmatrix} 209.6265 \\ 384.8318 \\ 247.7736 \\ 354.3515 \end{bmatrix} kV.$$

b. The objective function computed at the estimate is:

$$J = \sum_{i=1}^{8} \left(\frac{z_i - h_i^T \hat{x}}{\sigma_i} \right)^2 = (Hx - z)^T W (Hx - z) = 0.1289407.$$

The confidence level computed from the statistics of the residuals with 4 degrees of freedom (8 measurements—4 states=4), is approximately 0.998.

c. The standard deviation of the state variables is computed as follows:

$$C_x = \left(H^T W H\right)^{-1} = \begin{bmatrix} 35.9361 & 0 & 35.5930 & 0 \\ 0 & 35.9361 & 0 & 35.5930 \\ 35.5930 & 0 & 35.9361 & 0 \\ 0 & 35.5930 & 0 & 35.9361 \end{bmatrix},$$

$$\sigma_{x_1} = \sigma_{x_2} = \sigma_{x_3} = \sigma_{x_4} = \sqrt{35.9361} = 5.9947\ kV.$$

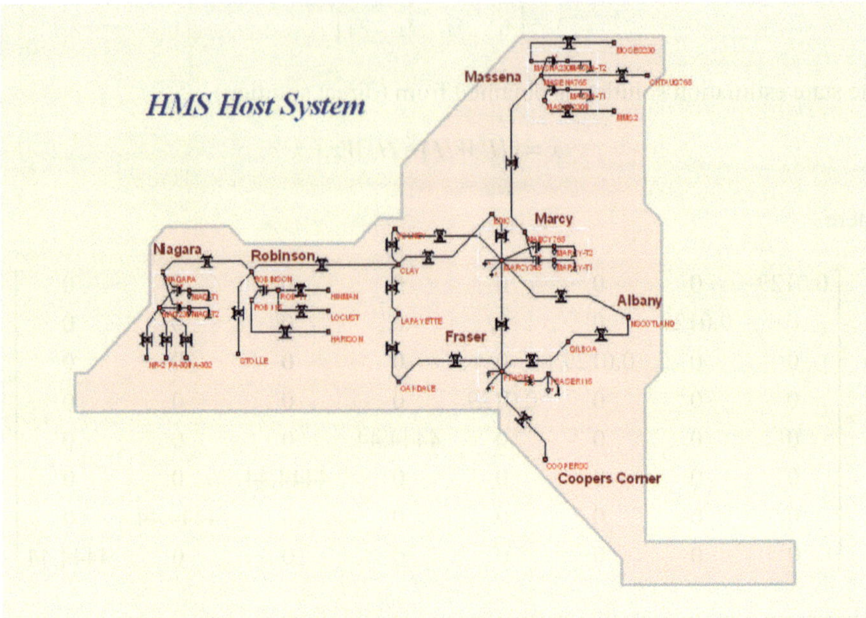

Fig. 2.9 System of the first three-phase linear state estimator installation (1993–1998)—Macrodyne PMUs installed at Niagara, Robinson, Massena, Marcy (2), Fraser, and Gilboa

Historical Note The Macrodyne PMU was introduced in January 1992 (Macrodyne PMU Model 1690). The first three-phase linear estimator implementation using the Macrodyne PMU was developed and installed on the NYPA system between 1993 and 1998 [5, 7]. The system consists of seven PMUs installed at several substations of the NYPA system shown in Fig. 2.9. The architecture of the system is shown in Fig. 2.10. Because one of the objectives of the project was to monitor harmonic flow in the system, the three-phase LSE was implemented for several harmonics, with the first harmonic being the fundamental, i.e., the standard power frequency LSE. The system works as follows. At each PMU location, the phasors for each phase (A, B, and C) are computed at fundamental and harmonic frequencies. A cut-off value of 0.5 % was used. Any harmonic phasor of magnitude below 0.5 % was ignored. Typically, this process would result in 1–5 harmonic phasors at each location, each harmonic phasor consisting of phase A, phase B, and phase C phasors. For example, at the FRASER location the harmonics were always present (there is a large static VAR compensator at this location). Then the fundamental frequency phasors are used for the three-phase LSE at the fundamental frequency, the third harmonic phasors are used for the three-phase LSE for the third harmonic, and so on. Details of the analytics are provided in [5].

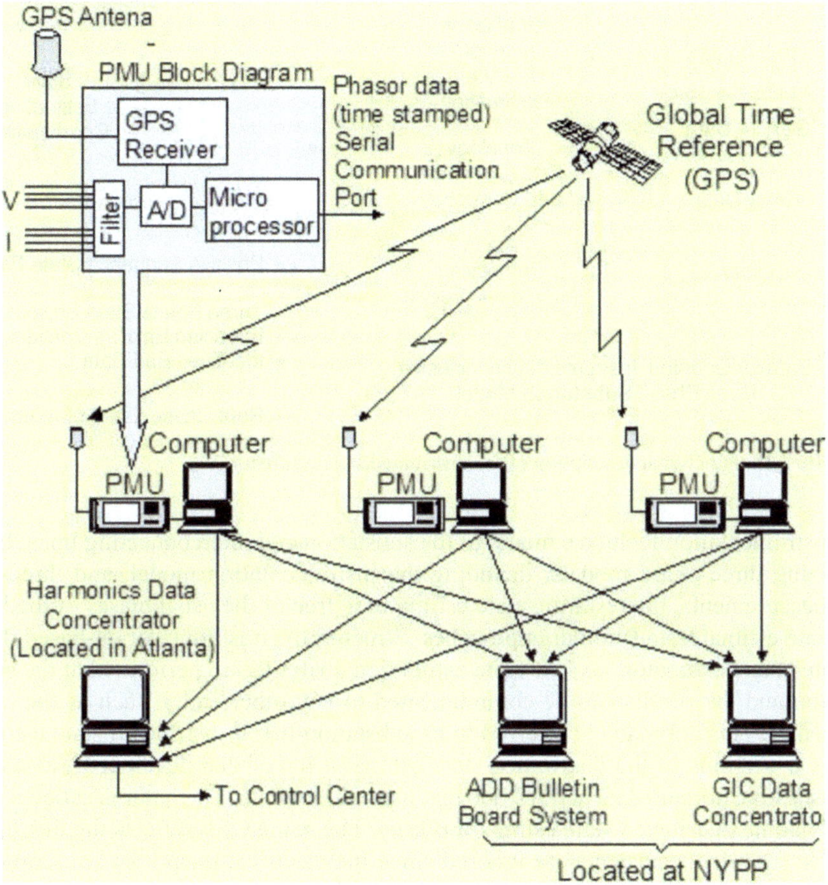

Fig. 2.10 Architecture of the first three-phase linear state estimator installation (1993–1998)

2.7 Substation-Based State Estimation

The advantage of utilizing synchrophasors for state estimation has been well understood. Another advantage of the introduction of synchrophasors that is less understood is that it made it possible to distribute the state estimation analytics. Specifically, the technology allows separating a large system into overlapping subsystems and performing the state estimation in each subsystem at exactly the same time. While the computations at each subsystem can be performed asynchronously, the measurement set used is synchronized with GPS accuracy among all subsystems. The results from each subsystem then can be combined to synthesize the state of the entire system. The distributed state estimation idea has been realized by the SuperCalibrator concept proposed by Meliopoulos and coworkers [9–11]. The SuperCalibrator is conceptually very simple and applies to each substation of a large system. It is simply a state estimator using a detailed three-phase breaker-oriented

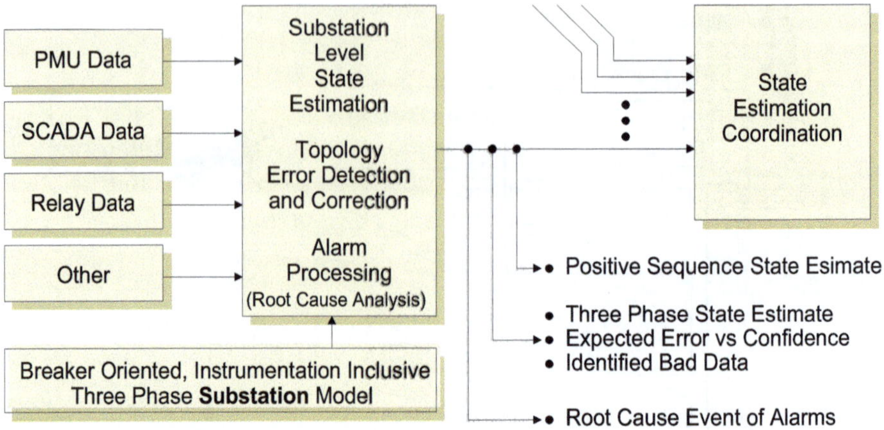

Fig. 2.11 Functional description of the distributed state estimator

instrumentation inclusive model of the substation and interconnecting lines. By utilizing three-phase models, including the instrumentation model, and three-phase measurements, the resulting state estimator is free of the usual biases of the legacy state estimator and similar approaches. Structurally, it is a substation-based distributed state estimator, i.e., the state estimation analytics are performed at the substation and the results can be communicated to any other entity, such as the control center. This approach is referred to as substation-based distributed state estimator. Two versions of the distributed state estimator have been developed: (a) one that uses time domain data and (b) one that uses phasor data. The latter is called a quasi-dynamic distributed state estimator (QSE). The name QSE reflects the fact that the fast electrical transients are ignored since the electrical quantities are represented with a phasor over a one-cycle time interval. However, other slower transients, such as machine oscillations are accounted for with a model that represents the slower dynamics with differential equations .

The functions of the QSE consists of (a) state estimation of the substation operating conditions based on three-phase, breaker-oriented, and instrumentation channel inclusive model, (b) bad data identification and rejection at the substation level, and (c) system-wide operating state construction. The functional description of the QSE is depicted in Fig. 2.11. A conceptual illustration of QSE is illustrated in Fig. 2.12. The use of the results from each substation for the construction of the entire system state estimate is illustrated in Fig. 2.13.

In subsequent paragraphs we introduce the analytics of the QSE.

2.8 QSE

This section describes the implementation of the distributed state estimator as an object-oriented and flexible program. There are two implementations depending on what dynamics are included in the system model. The two implementations are

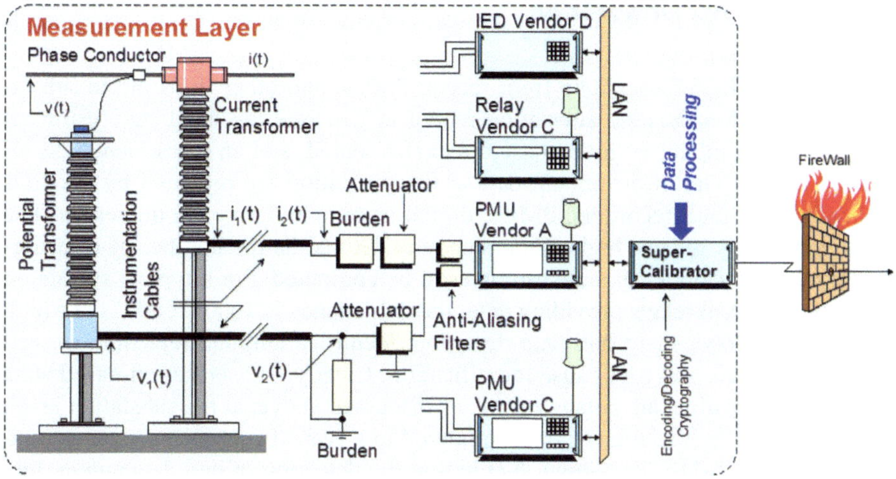

Fig. 2.12 Conceptual illustration of the QSE (SuperCalibrator)

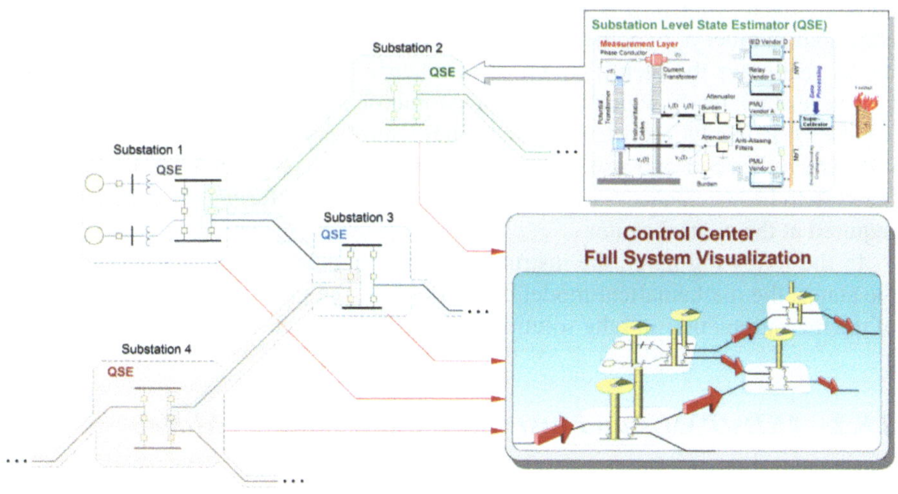

Fig. 2.13 Construction of the system-wide state estimate by synthesis of the substation state estimates

named quasi-dynamic state estimator (QSE) and domain dynamic state estimator (DSE). QSE models the electro-mechanical transients of the power system while the electrical transients are neglected. The DSE (time-domain) models both electrical and electromechanical dynamics. The focus of this book is on the QSE implementation. A brief description of the object-oriented approach for QSE is described next.

2.8.1 General Description

The distributed QSE uses data from relays, PMUs, and other IEDs in the substation along with a detailed substation model to generate the local state estimate. The substation model is three-phase, breaker-oriented, and an instrumentation inclusive model. The measurement data at the substation are provided by all IEDs (relays, digital fault recorders, PMUs, digital meters) and they are numerous. The data redundancy is very high. Redundancy is defined as the number of measurements over the number of state estimates to be computed. For a typical substation with all numerical relays providing data, the redundancy is 1000–2000 %. The high redundancy allows sharp bad data detection, identification, and rejection, as well as alarm analysis and root cause identification. Clearly, the substation-based state estimator has distinct advantages. First, as discussed above, at the substation level, there is greater redundancy of data compared to a typical centralized state estimator based on SCADA data alone, facilitating the detection of bad data and system topology errors. Second, the state estimator problem is much smaller in size and therefore powerful hypothesis testing methods are efficiently applied for both bad data and topology errors identification, i.e., without substantial deterioration of the computational efficiency, which is not the case in a centralized approach due to an enormous number of hypotheses. Third, the use of the three-phase, breaker-oriented model facilitates the identification of symmetric and asymmetric topology errors (one pole stuck, etc.), which traditional symmetric state estimators cannot identify.

The QSE results are transmitted to the control center where the system-wide state is synthesized and used for displays and other applications. Note that after the system has been successfully installed and tested, no further data processing is required at the control center.

In the following parts, the instrumentation channel modeling, the definition of the states, the mathematical model of the devices, the general model (object form) of the measurements, and the solution algorithm are provided.

2.8.2 IEDs/PMUs and Instrumentation Channel Model

An important issue for the formulation of the QSE is the accuracy of the available data. Specifically, PMUs are in general higher precision equipment as compared to typical SCADA systems or relays. PMUs provide measurements that are time tagged with time precision better than 1 μs (resulting in phase precision of 0.02 degrees for 60 Hz systems) and magnitude accuracy that is better than 0.1 %. This potential performance is not achieved in an actual field installation because of two reasons: (a) different vendors use different design approaches that result in variable performance among vendors, for example, use of multiplexing among channels or variable time latencies among manufacturers result in timing errors much greater than 1 μs, and (b) GPS-synchronized equipment receives inputs from instrument transformers, control cables, and associated infrastructure, which introduce magnitude and phase errors that are much greater than the accuracy of a typical PMU. These errors are referred to as the instrumentation channel error. The end result is that “raw” phasor data from different vendors cannot be used as highly accurate data.

Conceptually, the overall precision issue can be resolved with sophisticated calibration methods. This approach is quite expensive and faces difficult technical problems. In the early 1990s, Meliopoulos directed a research project in which calibration procedures were developed for selected NYPA’s high voltage instrument transformers [15]. While this approach is fine for a research project, it is economically prohibitive for system-wide error quantification and correction. An alternative approach is to utilize appropriate filtering techniques for the purpose of correcting the magnitude and phase errors, assuming that the characteristics of the various GPS-synchronized pieces of equipment are known and the instrumentation feeding this equipment is also known. This is done in the QSE approach by simply including the instrumentation model in the overall state estimation algorithm. Specifically, the models of the IEDs, PMUs, and the associated instrumentation channel model are all integrated into a single model that provides the transfer function from the high voltage side to the output of the IEDs and PMUs. This model is integrated with the three-phase, breaker-oriented model described earlier. The end result is an accurate representation of the physical system by which the measurements are taken. Details of the modeling of the instrumentation channels can be found in [16].

2.8.3 Definition of States

The state of the QSE is defined as the minimum information that completely describes the operating conditions of the substation model. Recall that the substation model comprises all equipments in the substation and all interconnecting lines. QSE uses voltage phasor states as well as other quantities such as speed of generator rotor, magnetic flux linkage, and mechanical power in order to represent the operating conditions of the system. The voltage states are called external state variables. For every three-phase bus i of the substation, the voltage state is defined as:

$$\tilde{V}_i = \left[\tilde{V}_{i,A}\tilde{V}_{i,B}\tilde{V}_{i,C}\tilde{V}_{i,N}\right]^T. \quad (2.5)$$

The state is extended to include internal states (algebraic or dynamic) of specific devices, e.g., the torque angle and the rotor speed of a generator or the magnetic flux linkage of a transformer. Furthermore, the system state also includes the states of the buses at the other ends of the lines/circuits connected to the substation. The conceptual illustration of the dynamic state estimator and the state definition is shown in Fig. 2.14.

All in all, the state in the QSE formulation is defined as:

$$X = \left[\tilde{V}, y(t), y(t_m)\right]^T,$$

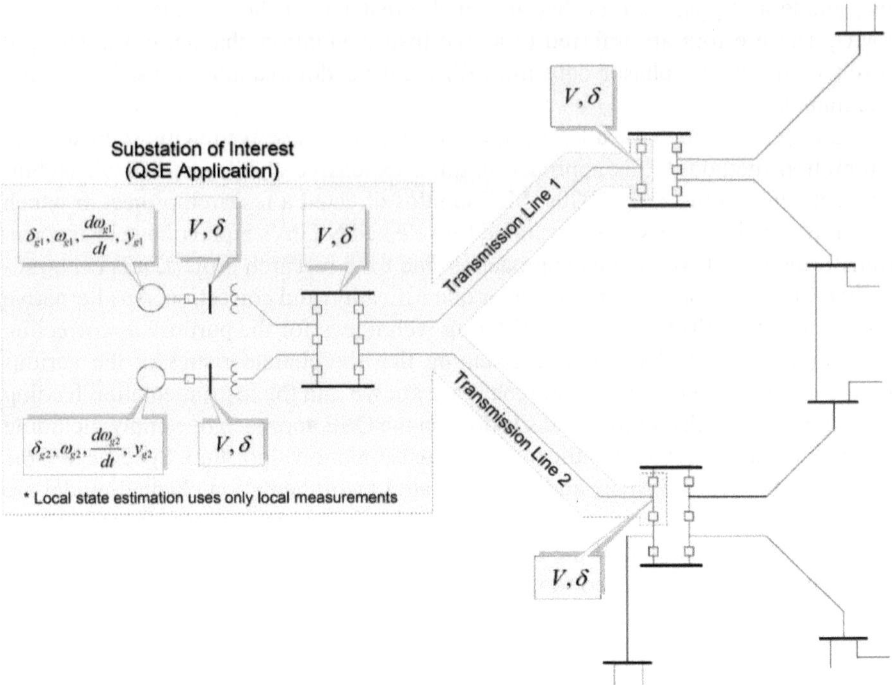

Fig. 2.14 QSE—definition of states for the substation model

where $\tilde{V}$ is the vector of voltage states at buses (phasor form) and $y(t)$, $y(t_m)$are the vectors of dynamic states at times t and t_m, respectively.

2.8.4 Quadratic Component Modeling and Quadratic Integration

In order for the estimation algorithm to be able to accurately capture the dynamic phenomena, the power system is modeled using a physically based, three-phase dynamic model with explicit representation of all devices of the system such as transmission lines, transformers, and generators. Fast electrical transients are neglected (i.e., electrical quantities are represented with phasors, which are functions of time), but slower electromechanical transients are modeled with differential equations. The states of the differential equations are added to the state of the state estimator. This approach results in a more realistic representation of the actual power system and the ability to estimate the dynamic and transient behavior of the power system. The end result is that a device is described by a set of differential and algebraic equations, as follows:

$$\frac{dx_1(t)}{dt} = f\left[x_1(t), x_2(t), t\right] \text{and } 0 = g\left[x_1(t), x_2(t), t\right], \tag{2.6}$$

where the $x_1(t)$ and $x_2(t)$are the dynamic and algebraic states of the system.

In general, the above model may be linear or nonlinear. Nonlinear models may have higher order nonlinearities. In case of higher order nonlinearities, additional state variables can be introduced to reduce the order of nonlinearity. As an example, consider a term x^3. This nonlinear term can be replaced with $x^3=xy$ where $y=x^2$. Note that the third order nonlinearity has been reduced to second order but the model now is augmented with one additional equation, i.e., $y=x^2$. We refer to this process as a quadratization procedure [13], [17]. The resulting equations from the quadratization are cast in the following form:

$$\frac{dx_1(t)}{dt} = A_1x_1(t)+A_2x_2(t)+A_3x_3(t)+C, \tag{2.7}$$

$$0 = E_1x_1(t)+E_2x_2(t)+E_3x_3(t)+\begin{bmatrix} [x_1(t),x_2(t),x_3(t)]F_1[x_1(t),x_2(t),x_3(t)]^T \\ \vdots \\ [x_1(t),x_2(t),x_3(t)]F_n[x_1(t),x_2(t),x_3(t)]^T \end{bmatrix}+D, \tag{2.8}$$

where $x_3(t)$ are additional algebraic (internal) states introduced for the quadratization of the model. Note that in the proposed formulation there are no nonlinearities in the dynamic part of the model. That is, all nonlinearities are moved to the algebraic part of the model through the introduction of additional appropriate state variables. Also note that the nonlinear algebraic equations are of degree no more than two (at most quadratic equations), which is also achieved by the introduction of additional appropriate state variables. Assuming that the dynamic functions vary quadratically within the integration step $[t\text{-}h,\ t]$, and upon quadratic integration [18], the dynamic model of the device is converted into a set of algebraic equations. The nonlinearities involved in these equations are no higher than quadratic. The involved variables $x_1(t)$, $x_2(t)$, and $x_3(t)$ are separated into through variables (electric currents at the terminals of the components), denoted by $i(t)$, across variables (voltage at the terminals of the component) denoted by $v(t)$ and the remaining variables (internal state variables) denoted by $y(t)$. Using this notation, the algebraic equations are written in the following form (details in [18]):

$$\begin{bmatrix} i(t) \\ 0 \\ i(t_m) \\ 0 \end{bmatrix} = Y_{eq}\begin{bmatrix} v(t) \\ y(t) \\ v(t_m) \\ y(t_m) \end{bmatrix} + \begin{bmatrix} \left[v^T(t)\,y^T(t)\,v^T(t_m)\,y^T(t_m)\right]\cdot F_{eq\_1}\cdot\begin{bmatrix} v(t) \\ y(t) \\ v(t_m) \\ y(t_m) \end{bmatrix} \\ \vdots \\ \left[v^T(t)\,y^T(t)\,v^T(t_m)\,y^T(t_m)\right]\cdot F_{eq\_n}\cdot\begin{bmatrix} v(t) \\ u(t) \\ v(t_m) \\ y(t_m) \end{bmatrix} \end{bmatrix} - B_{eq}, \tag{2.9}$$

where subscripts t and t_{m} denote the time point t and the middle time point between t and t–h.

$i(t)$ the through variables of the device model
$v(t)$ the external states of the device model
$y(t)$ the internal states of the device model
Y_{eq} matrix defining the linear part of the device model
F_{eq} matrix defining the quadratic part of the device model
B_{eq} constant vector of the device model

We will refer to this formulation as the quadratized algebraic companion form (QACF) of the device. This form contains the equations for the through variables of the model (in our case terminal currents for each device terminal) along with the additional model equations. The equations of the through variables are referred to as external or interface equations, while the rest of the model equations are internal. Note that QACF is the three-phase and fully asymmetrical dynamic model of the device. This model is used in the formulation of the state estimation measurement model, as explained in detail later.

There are two main advantages to this formulation. First, it gives the generalized and abstract model for any component of the power system, which is independent of the type of the device and is suitable for implementation of an object-oriented algorithm. Moreover, the resulting equations of the system model are either linear or quadratic; thus, application of weighted least square method (which uses Newton's method) is ideally suited. As a result, such a quadratic formulation of a system model provides faster convergence of the state estimation algorithm and eliminates convergence issues.

2.8.5 Measurement Set and Measurement Model

The measurement set comprises of (a) GPS-synchronized voltage and current phasor measurements and (b) traditional, nonsynchronized measurements (voltage magnitude, line current magnitude, and any other measurements available from any IEDs in the substation). In addition, to the actual measurements the approach is facilitated by a number of derived, virtual, and pseudo measurements that are properly introduced.

In the proposed formulation, all the measurements are expressed with a generalized formula (see (Eq. (2.9)) as nonlinear functions in terms of the states with order at most quadratic, thus allowing implementation of the object-oriented approach. The generalized form of a measurement is as follows, and it is referred to as the QACF model of the measurement:

$$\begin{aligned} z_k(t) &= \sum_i a_{i,t}^k \cdot x_i(t) + \sum_i a_{i,im}^k \cdot x_i(t_m) + \sum_{i,j} b_{i,j,t}^k \cdot x(t) \cdot x_j(t) \\ &\quad + \sum_{i,j} b_{i,j,tm}^k \cdot x_i(t_m) \cdot x_j(t_m) + c_k(t) + \eta_k, \text{ and } \sigma_{n_k} = \sqrt{\mathrm{Var}(n_k)}, \end{aligned} \tag{2.10}$$

where z_k is the measured value

c_k is the constant term
$a_{k,i}$ are the linear coefficients
$b_{k,i}$ are the nonlinear coefficients
η_k is the error term.

We can classify the measurements into five categories:

- Actual across measurements
- Pseudo across measurement
- Actual through measurements
- Virtual (noiseless) through measurements
- Pseudo through measurements
- Derived measurements

Actual across measurements are real measurements that are not related to a device, e.g., a voltage measurement between two nodes (GPS-synchronized or nonsynchronized).

Pseudo across measurements are not real measurements, and are also not related to a device. For example, if only the voltage phasor measurements of phase A is available, pseudo measurements for phase B and C with the same magnitude and a phase difference of +/− 120 can be added accordingly. Since these are not actual measurements, there is greater uncertainty, and a larger measurement error is assigned to these measurements.

Actual through measurement is a real measurement related to a device, e.g., current measurement at a terminal of a transmission line (GPS-synchronized or nonsynchronized).

Virtual through measurement is not a real measurement; it represents a quantity that may be known from a physical law, e.g., the sum of the currents in a node must equal zero. The zero represents a measurement with zero error. Similarly, other physical laws (model equations) that a device must satisfy represent virtual measurements.

Pseudo through measurements are not real measurements, and they are related to a device. For example, if only the current phasor measurements of phase A of a device is only available, pseudo measurements for phase B and C with the same magnitude and a phase difference of +/− 120 can be added accordingly.

Finally, derived measurements are through measurements derived from either a device or node connectivity analysis. For example, assume a current measurement that is taken at node 1 of a single-phase capacitor; then, a derived current measurement for node 2 with the opposite value compared to the original measurement can be created and used in the estimation. Another example of a derived measurement is: Assume a node where three branches are connected and there are GPS-synchronized current measurements for two of the devices. Then a derived measurement can be created for the current in the third device with the value of the measurement being the negative of the sum of the two original measurements. Details on the types of measurements and their models can be found in [11] and [12].

2.8.6 State Estimation Algorithm

As described earlier, each measurement is related to the model via an algebraic linear or quadratic equation:

$$z_m = h(x, y) + \eta_m, \tag{2.11}$$

where x, y are the external and internal states respectively and the function h is at maximum quadratic function in terms of x and y.

The state estimation is solved using the weighted least squares approach. The problem is formulated as follows:

$$\text{Minimize } J(x, y) = \eta^T W \eta, \tag{2.12}$$

where $\eta = z - h(x, y)$ and W is a diagonal matrix, whose nonzero entries are equal to the inverse of the variance of the measurement errors:

$$W = diag\left(\frac{1}{\sigma_v^2}\right). \tag{2.13}$$

The best estimate of the system state is obtained from the Gauss–Newton iterative algorithm:

$$\hat{x}^{j+1} = \hat{x}^j - \left(H^T W H\right)^{-1} H^T W\left[h\left(\hat{x}^j\right) - z_a\right], \tag{2.14}$$

where $\tilde{x}$ refers to the best estimate of the state vector, and H is the Jacobian matrix of the measurement equations.

Note that if all the models are linear and all measurements are GPS-synchronized measurements, algorithm Eq. (2.14) becomes a direct solution (one iteration), i.e., it becomes a LSE.

2.8.7 Performance Metrics

The accuracy of the state estimate is quantified via the standard chi-square test and the statistical properties of the state estimator. The chi-square test defines the probability that the distribution of the measurement errors is within the expected bounds. Given the number of measurements m and the number of states n, the degrees of freedom can simply be calculated as $v=m-n$. Calculating the value ζ of the objective function based on the state estimates, the estimation confidence level is given by the probability:

$$\Pr\left[\chi^2 \geq \zeta\right] = 1.0 - \Pr\left[\chi^2 \leq \zeta\right] = 1.0 - \Pr(\zeta, v). \tag{2.15}$$

The chi-square test is utilized to provide the probability that the error of the measurements is within the range of errors of the IED with which the measurements were obtained. Since the various data acquisition devices in any substation may have different accuracy, and for the purpose of quantifying the performance of the state estimator with one single variable, a normalization constant k is introduced. The variable k is defined as follows: if it is 1.0 then the standard deviation of each measurement is equal to the accuracy of the meter with which this measurement was obtained. If different than 1.0 then the standard deviation of the measurement error equals the accuracy of the meter times k. Mathematically, this means that the standard deviation of the error of a device i is expressed as:

$$\sigma_i' = k\sigma_i.$$

Now the probability can be expressed in terms of the variable k:

$$\Pr\left[\chi^2 \geq \zeta^0 / k^2\right] = 1.0 - \Pr\left[\chi^2 \leq \zeta^0 / k^2\right] = 1.0 - \Pr\left(\zeta^0 / k^2, v\right),$$

where

$$\zeta^0 = \sum_{i=1}^{m} \left(\frac{h_i(x) - z_i}{\sigma_i}\right)^2 \text{ and } \zeta^0 / k = \sum_{i=1}^{m} \left(\frac{h_i(x) - z_i}{k\sigma_i}\right)^2.$$

The introduction of the variable k allows the characterization of the accuracy of the estimated states with only one variable, i.e., the variable k.

Other performance metrics can be also computed. For example, the accuracy of the solution (accuracy of the estimated state variables) is computed via the covariance (or information) matrix. The covariance matrix of the state is defined as:

$$C_x = E\left[(\hat{x} - \overline{x})(\hat{x} - \overline{x})^T\right], \tag{2.16}$$

where $\overline{x}$ denotes the true state value and $\hat{x}$ the estimated value, and is computed as

$$C_x = \left(H^T W H\right)^{-1}. \tag{2.17}$$

The above matrix is also known as the information matrix. Once the information matrix has been computed, the standard deviation of a state variable x_i is given with

$$\sigma_{xi} = \sqrt{C_x(i,i)}, \tag{2.18}$$

where $C_x(i, i)$ is the ith diagonal entry of the C_x.

The estimates of the measurements can also be computed as:

$$\hat{b} = h(\hat{x}, \hat{y}) \tag{2.19}$$

and their covariance matrix is:

$$Cov\left(\hat{b}\right) = H\left(H^T W H\right)^{-1} H^T. \tag{2.20}$$

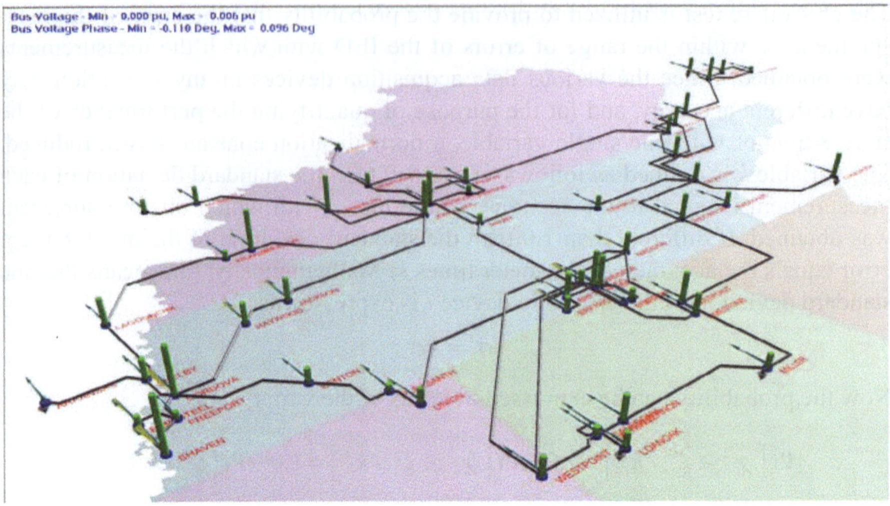

Fig. 2.15 3D rendering of the voltage magnitude and phase angle residuals

2.8.8 *Visualizations*

The results of the state estimator can be visualized via a variety of displays. One useful display is the illustration of residuals, i.e., the differences between measured values and estimated values. Figure 2.15 illustrates such visualization. The visualization shows the voltage magnitude residuals as tubes, where the length of the tube is proportional to the size of the residual. It also shows the residuals of the phase angle as arcs. The size of the arc is proportional to the phase angle residual. In these visualizations, and because the residuals are typically very small, it is customary to display magnified values of the residuals. As an example, the residuals illustrated in Fig. 2.15 are magnified by a factor of 50. In addition, in the upper left corner the largest values of the actual residual values are listed.

Visualizations are also very important for investigating various options of state estimators (e.g., effect of model selection on the overall performance of the state estimator), the effect of missing measurements, as well as identification of bad measurements. Example results are presented below for these two examples.

Effect of non-three-phase measurements As discussed, the measurements for the substation-based state estimator are coming from devices such as relays, fault recorders, and PMUs. All of these devices can provide three-phase measurements, e.g., three phasors for the voltages at each bus, or three phasors for the current at a circuit. The PMUs may be also set up to provide only the positive sequence quantities, or all phase quantities. To assess the impact of measurements on the performance of the state estimator, a numerical experiment was performed as follows: Only the phase A measurements were used (both voltages and currents). For

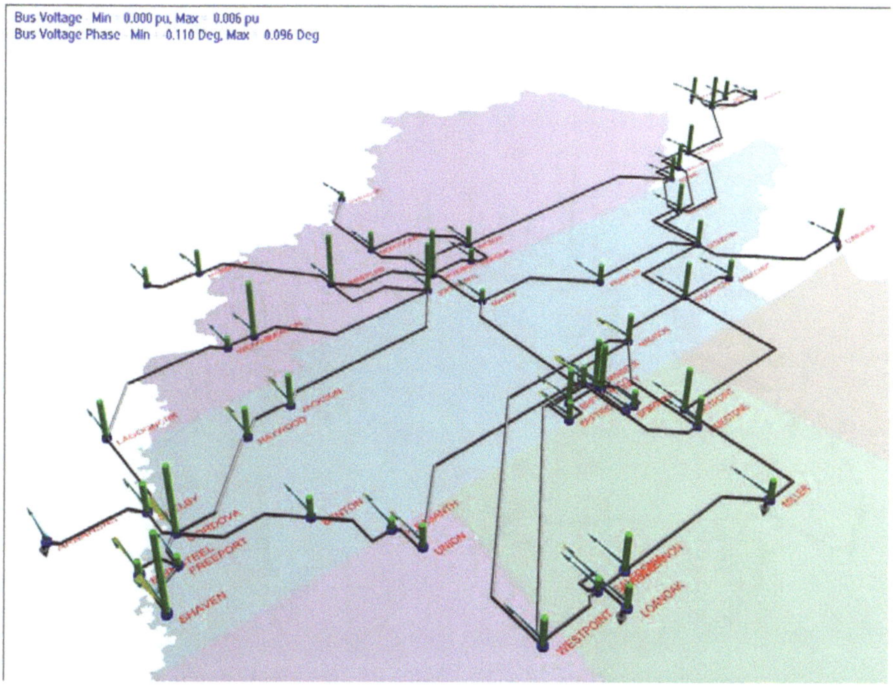

Fig. 2.16 3D rendering of the voltage magnitude and phase angle residuals of phase A voltages—only phase A measurements were used in the state estimator

observability, pseudo measurements were introduced for the other phases by simply taking the phase A measurements and applying a 120- and 240-degree phase shift for the phases B and C, respectively. This was done for both voltage and current measurements. The results of the state estimator in this case are displayed in Figs. 2.16, 2.17, and 2.18 for phases A, B, and C, respectively. The results are presented in terms of the magnitude and phase residuals. The displayed residuals are amplified by a factor of 50; the actual maximum values of the residuals are also given for reference. Note in Fig. 2.16 the residuals for phase A are relatively very small. In Fig. 2.17, the residuals for phase B are much larger and similarly for phase C results in Fig. 2.18. It is clear that without actual phase B and phase C measurements, the residuals for these phases are relatively large. The magnitude depends on the system imbalance. This system has relatively small imbalance, but yet the results clearly show a deterioration of the state estimator performance for the unmeasured phases.

Identification of Bad Measurements Another area where visualization methods are useful is the area of bad measurement identification. For the centralized state estimator this is a very time-consuming task as typically hypothesis testing is performed to identify bad measurements. In the case of the centralized state estimator, another issue that makes the task difficult is the low redundancy level of measurements (typically 200–300 %). The substation-based state estimator has normally

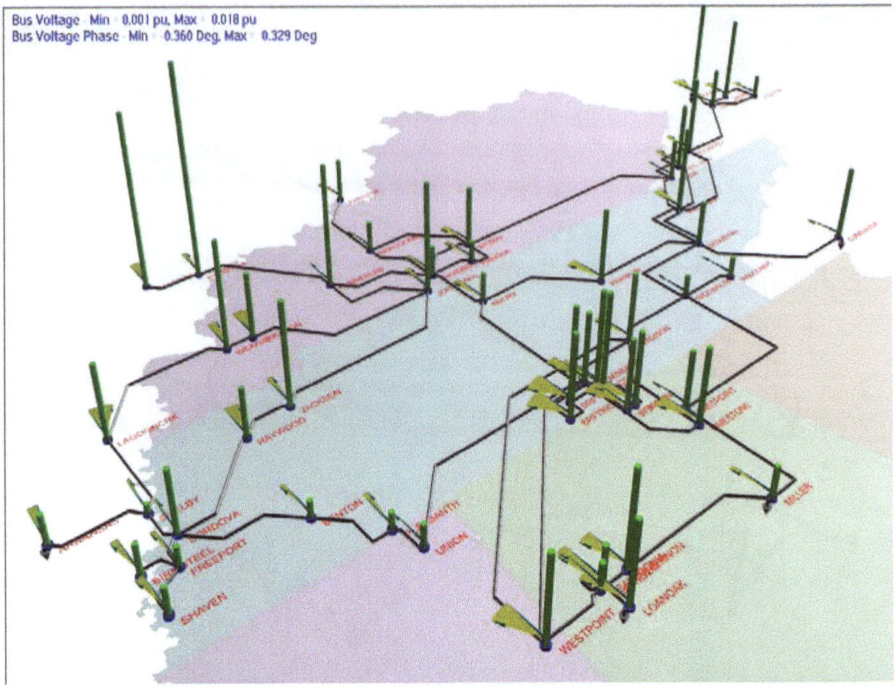

Fig. 2.17 3D rendering of the voltage magnitude and phase angle residuals of phase B voltages—only phase A measurements were used in the state estimator

much higher redundancy (typically 1000–2000 %) and it is much easier to identify the bad measurements. Visualization helps to identify bad measurements via inspection. Figure 2.19 shows the visualization of the voltage residuals, both phase and magnitude, after a bad measurement of 50 MW, error was introduced in the flow of a line. Note that the phase residual has become relatively large at the two end of the line. The largest value of the residuals is 0.672 degrees. Figure 2.19 shows the residual magnified by a factor of 200.

2.9 Summary and Conclusions

In this chapter, the applications of PMU measurements were examined. Specifically, the focus was on state estimation. State estimation is a basic application that provides the real time model of the system and quantifies the accuracy of this model. Thus, state estimation enables a large number of other applications that require the real time model of the system, such as security assessment, voltage and reactive power scheduling, and other applications. PMU measurements have changed our approaches to state estimation as new PMU-based state estimators have emerged

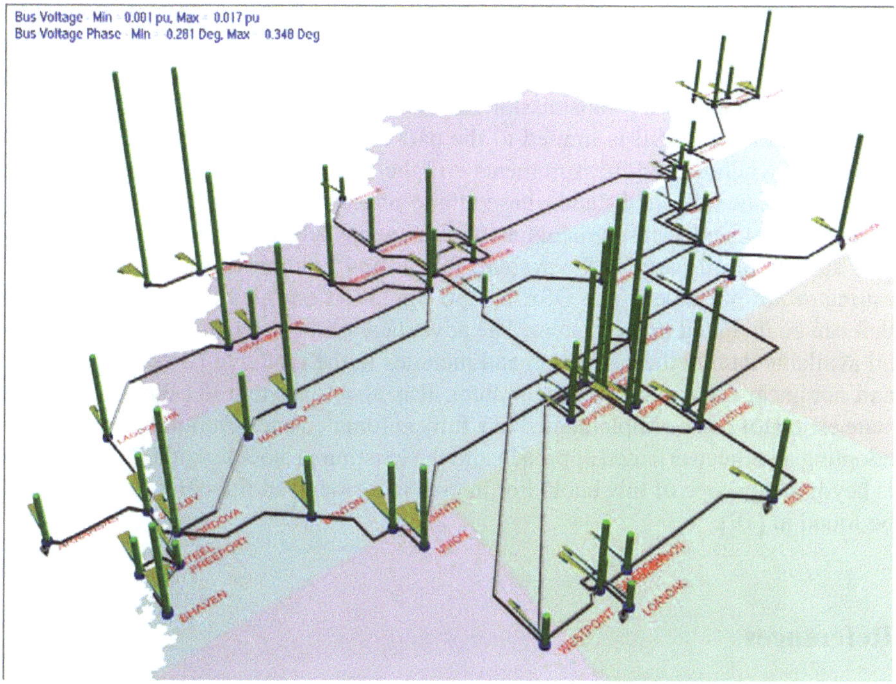

Fig. 2.18 3D rendering of the voltage magnitude and phase angle residuals of phase C voltages—only phase A measurements were used in the state estimator

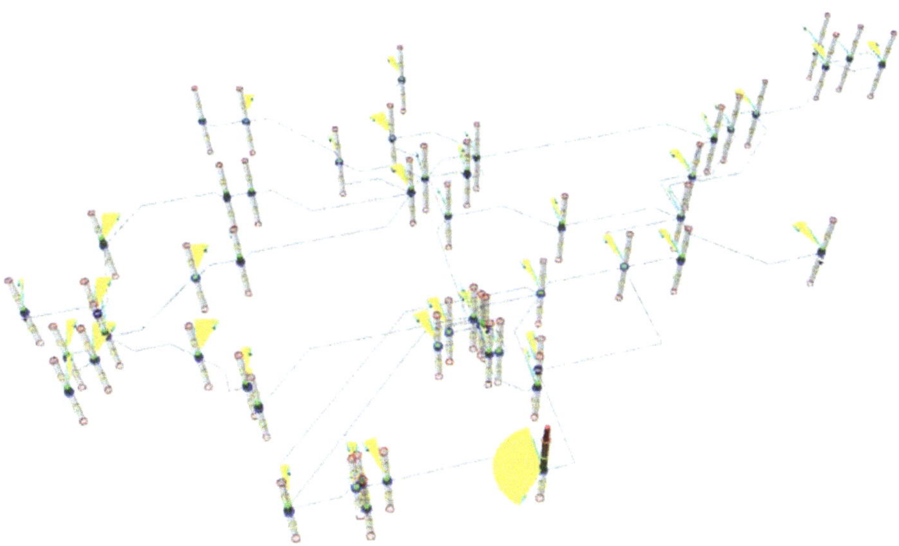

Fig. 2.19 3D rendering of the voltage magnitude and phase angle residuals of phase A voltages—a bad measurement with 50 MW error was introduced in the nearby line

with tremendous capabilities. The legacy state estimator that is based on non GPS-synchronized data was first introduced. The three-phase LSE that is dependent 100% on GPS-synchronized measurements and confined on the part of the system with linear devices (i.e., transmission lines) was also discussed. It is important to emphasize that the LSE is limited to the part of the system that is fully equipped with GPS-synchronized measurements and the components of the system are linear. Finally, the QSE (substation-based three-phase state estimator) uses data from GPS-synchronized measurements as well as nonsynchronized measurements and it is applied to all parts of the system. Linear and nonlinear versions of the QSE estimator are presented. The QSE has no limitation on the type of system model that can be included in the system. The advantage of this estimator is that it utilizes all available data in the substation and includes in the model to be estimated linear and nonlinear components of the system. It is also important to point out that the state estimator can be implemented in a fully automated and autonomous manner by adopting an object-oriented approach and the existing protocol standards. This issue is beyond the scope of this book. For interested readers, additional information can be found in [19].

References

1. IEEE Standard C37.118, IEEE standard for synchrophasors for power systems, 2011
2. A.P. Sakis Meliopoulos, G. Stefopoulos, Characterization of state estimation biases. Probab. Eng. Inform. Sci. **20**(1), 157–174 (January 2006)
3. J.S. Thorp, A.G. Phadke, K.J. Karimi, Real time voltage-phasor measurements for static state estimation. IEEE Trans. Power Appar. Syst. **PAS-104**(11), 3098–3106 (1985)
4. A.G. Phadke, J.S. Thorp, K.J. Karimi, State estimation with phasor measurements. *Proceedings of 1985 PICA Conference*, pp. 142–147, 1985.
5. A.P. Sakis Meliopoulos, F. Zhang, S. Zelingher, Power system harmonic state estimation. IEEE Trans. Power Syst. **9**(3), 1701–1709 (1994)
6. A.P. Sakis Meliopoulos, F. Zhang, Multiphase power flow and state estimation for power distribution systems. IEEE Trans. Power Syst. **11**(2), 939–946 (1996)
7. B. Fardanesh, S. Zelingher, A.P. Sakis Meliopoulos, G. Cokkinides, J. Ingleson, Multifunctional synchronized measurement network. IEEE Comput. Appl. Power **11**(1), 26–30 (1998)
8. A.A. Jun Zhu, M. Rice, G.T. Heydt, S. Meliopoulos, *Enhanced State Estimators Arizona State University* (PSERC Publication, 2006)
9. A.P. Sakis Meliopoulos, G.J. Cokkinides, C. Hedrington, T. Conrad, The SuperCalibrator—a fully distributed state estimator. *Proceedings of the IEEE-PES 2010 general meeting*, Minneapolis, MN, 26–29 July 2010
10. S. Mohagheghi, G.J. Cokkinides, A.P. Sakis Meliopoulos, Distributed state estimation based on the SuperCalibrator concept—laboratory implementation, *Proceeding of IREP 2007, Symposium on Bulk Power System Dynamics and Control,* 2007
11. A.P.S. Meliopoulos, G.J. Cokkinides, F. Galvan, B. Fardanesh, Distributed state estimator advances and demonstration. *Proceedings of the 41st annual Hawaii international conference on system sciences*, pp. 163–163, 2008
12. A.P.S. Meliopoulos, G.J. Cokkinides, F. Galvan, B. Fardanesh, P. Myrda, Advances in the SuperCalibrator concept—practical implementations. *Presented at the proceedings of the 40th annual Hawaii international conference on system sciences*, 2007

13. S. Meliopoulos, G.J. Cokkinides, G. Stefopoulos, T. Conrad, C. Hedrington, Distributed state estimator via the SuperCalibrator approach. Protection, Automation and Control World, pp. 38–44, Autumn 2009
14. S. Meliopoulos, G.J. Cokkinides, F. Galvan, B. Fardanesh, P. Myrda, Delivering accurate and timely data to all: Model based substation automation applications for advanced data availability. IEEE Power Energy Mag. **5**(3), 74–86 (2007)
15. A.P. Meliopoulos, F. Zhang, S. Zelingher, G. Stillmam, G.J. Cokkinides, L. Coffeen, R. Burnett, J. McBride, Transmission level instrument transformers and transient event recorders characterization for harmonic measurements. IEEE Trans. Power Deliv. **8**(3), 1507–1517 (1993)
16. A.P. Sakis Meliopoulos, G.J. Cokkinides, Visualization and animation of instrumentation channel effects on DFR data accuracy. *Proceedings of the 2002 Georgia tech fault and disturbance analysis conference*, Atlanta, Georgia, 29–30 April 2002
17. A.P. Meliopoulos, G.J. Cokkinides, G.K. Stefopoulos, Improved numerical integration method for power/power electronic systems based on three-point collocation. *Proceeding of the 44th IEEE conference on decision and control, and European control conference*, pp. 6780–6787, 2005
18. A.P.S. Meliopoulos, G.J. Cokkinides, G. Stefopoulos, Quadratic integration method. *Proceeding of international power system transients conference*, Montreal, 2005
19. S. Choi, B. Kim, G. Cokkinides, A.P. Sakis Meliopoulos, Feasibility study: autonomous state estimation in distribution systems. IEEE Trans. Power Syst. **26**(4), 2109–2117 (2011)

Chapter 3
Real Time Stability Monitoring

3.1 Stability Phenomena in Power Systems

Power system operation is undergoing major technological advances with many new installations of synchrophasors all across the North American grid as well as in power systems all over the world. Recent initiatives by the US federal government in the general area of smart grid technology are contributing to installations of synchrophasor monitoring systems in many utilities in the North American power grid. The industry–university collaborative organization, North American Synchrophasor Initiative (NASPI), is serving as a major coordinating group in leading the recent efforts in this direction. Synchrophasor measurements together with modern communication technology facilitate the monitoring of the current state of the wide-area power system in near real time. Recent research aims to exploit the availability of such wide-area synchronized measurements from across the system into developing new real time tools for power system operation.

Power system operation is normally required to meet the following four operational reliability properties [1]: (1) acceptability or viability (all voltages and currents stay within specified acceptable tolerances), (2) small-signal stability (system dynamics is able to damp out all small-scale disturbances), (3) transient stability (system dynamics can recover from all credible contingencies), and (4) voltage stability (system dynamics renders operation at nominal viable voltages without degenerating into voltage collapses or voltage declines). More extensive coverage of power system stability concepts can be seen in standard textbooks such as [2] and [3]. Sections 3.2–3.4 of this chapter will highlight operational tools related to the three properties, namely, small-signal stability, transient stability, and voltage stability. The discussions in this chapter are aimed at real-time algorithms that are purely based on wide-area synchronous measurements without knowledge of the underlying nonlinear power system model. Chapter 4, on the other hand, will discuss tools that combine off-line model-based analysis with real time wide-area synchronized measurements for hybrid online approaches.

M. Kezunovic et al., *Application of Time-Synchronized Measurements in Power System Transmission Networks*, Power Electronics and Power Systems,
DOI 10.1007/978-3-319-06218-1_3,

3.1.1 Voltage Stability Phenomena

Voltage stability refers to the ability of a power system to maintain its bus voltages at acceptable voltage levels while withstanding (a) variations in system loads and tie-line flows and (b) loss of components such as generators and transmission lines. It is well known that the problem of voltage stability is closely related to the problem of maintaining adequate reactive power support. Reactive power losses over typical power system transmission lines are relatively much higher compared to respective active power losses under normal and stressed operating conditions. Therefore, unlike active power, reactive power does not travel far in terms of electrical distances from reactive power sources. However, reactive power demands can become unusually high in some portions of the power grid for a variety of reasons such as (a) heavy summer loads with large air conditioning load demands, (b) sudden loss of critical transmission support, and (c) loss of key volt ampere reactive (VAr) support devices such as generators and static VAr compensators. Under these circumstances, if there is not enough local VAr support to mitigate the increased VAr demand, voltage collapse can occur. For instance, the July 2, 1996 Western American blackout was caused by lack of adequate VAr support in the Idaho portion of the western grid [4]. This was in turn related to unusually high VAr demands from heavy air conditioning loads on that day in the Idaho region and from atypical transmission tie-line flows in the neighboring power system.

The time-plots of recorded bus voltage magnitudes at two critical buses, namely, from Boise 230 kV bus in Idaho region, and from Malin 500 kV bus on California–Oregon tie-lines, are shown in Fig. 3.1. The event started with a routine fault on one of the transmission lines near the Jim Bridger generating units in Wyoming, denoted by time 0 in the time-plots of Fig. 3.1. There was a complex sequence of line and generator trippings during the initial 24 s as seen in Fig. 3.1; details can be seen in [4]. The time-plots in Fig. 3.1 show that the Boise bus voltage collapsed from about 200 kV to about 100 kV in a few seconds after 25 s from the start of the event. Shortly afterwards, the Malin bus voltage on the crucial tie-lines connecting Oregon and California also collapsed from about 530 kV down to about 300 kV, which led to the tripping of the California–Oregon tie-lines and the subsequent blackout in the western power grid.

Postmortem analysis of power-flow files modeling the system conditions during the disturbance showed that the system was operating outside normal power-flow solvability limits during the event [4]. For instance, the QV plot for Boise 230 kV bus is shown in Fig. 3.2. Here for generating the QV plot, a fictitious VAr demand is added to Boise 230 kV bus for the power-flow conditions that correspond to the system conditions just after the initial set of trippings at around 5 s, as seen in the plot of Fig. 3.1. Then, the power-flow solutions are tracked while varying the reactive power demand at the Boise bus. The QV plot in Fig. 3.2 shows the respective values of the Boise 230 kV bus voltage for different VAr demands. It is important to note that the power-flow could be solved only by adding a minimum of 140 MVAr VAr support at the Boise load bus, which implies that the VAr margin was *negative*

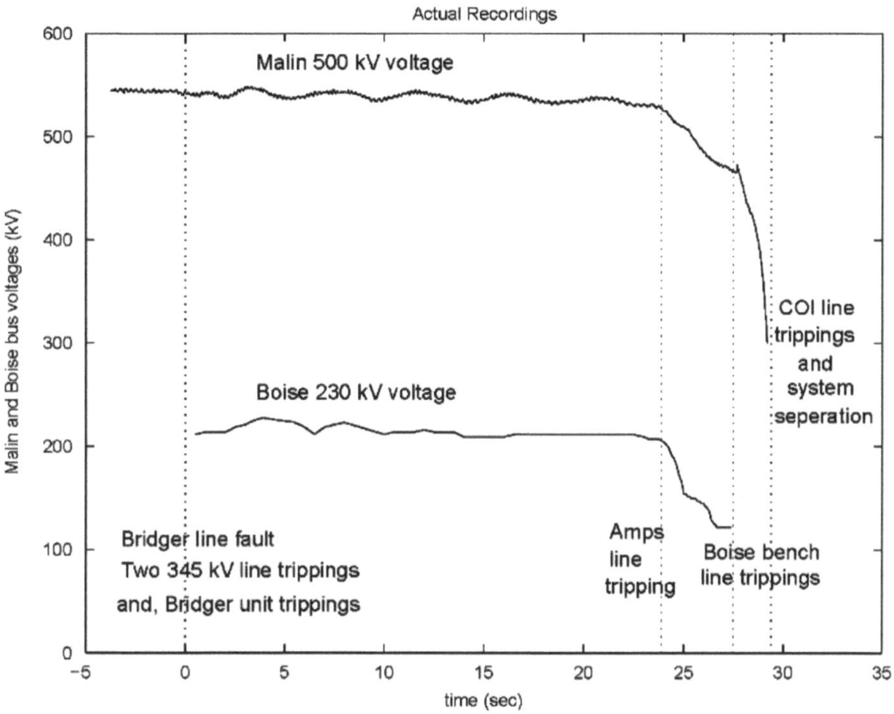

Fig. 3.1 Recorded bus voltage magnitudes during July 2, 1996 blackout [4]. *COI* California–Oregon Intertie

at about −140 MVAr at the Boise 230 kV bus at that time [4]. In other words, the system was operating outside power-flow solvability limits for the time period after 5 s, which led to tripping of other components and the subsequent voltage collapse within 30 s, as shown in Fig. 3.1. This example shows the importance of maintaining adequate VAr margins in the power system at all load buses and the need to monitor them continuously.

3.1.2 Small-Signal Stability Phenomena

Power systems are continuously subjected to small-scale and large-scale disturbances. Small disturbances are caused by random load fluctuations throughout the power grid as well as by variations in renewable power generation sources increasingly present in modern power systems. On the other hand, tripping of major transmission lines, generators, and other power system equipment such as transformers can lead to large disturbances. From an operational reliability point of view, the power system has to be able to withstand such disturbances and continue to operate nominally irrespective of whether the disturbances are small or large. However, the analysis of

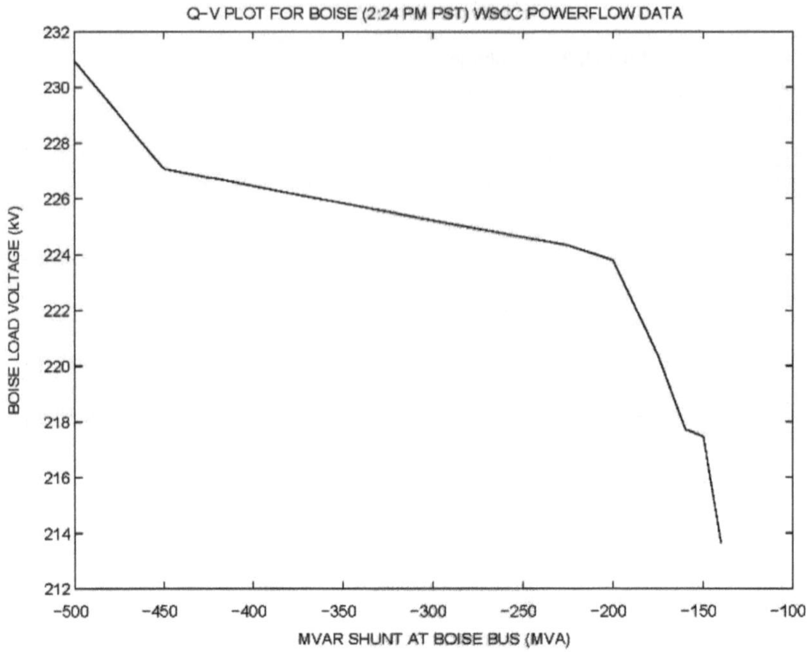

Fig. 3.2 Post-transient QV plot at Boise 230 kV load bus [4]

system behavior following disturbances is very different in the context of dynamic analysis when small disturbances are compared to large contingencies.

The power system behavior can be modeled by a large set of nonlinear differential algebraic equations [1–3]. For small disturbances, the nonlinear response can be linearized [3] and principles of linear system theory can be applied in understanding the dynamic phenomena, e.g., by computing the eigenvalues of the underlying linearized dynamic model or the "oscillatory modes" of a power system [3]. On the other hand, nonlinearity must be preserved in analyzing the post-disturbance responses of a power system following large disturbances [3].

Assuming the power system is operating at a steady state equilibrium condition, small-signal stability refers to the system's ability to return to the equilibrium condition after any small disturbance. In nonlinear system theory, the concept is referred to as local asymptotic stability [1]. Theoretically, when all the eigenvalues of the linearized Jacobian matrix (evaluated at the equilibrium point) have negative real parts, the equilibrium can be proved to be locally asymptotically stable [2]. Conversely, when some eigenvalues have positive real parts for whatever reason, the equilibrium becomes small-signal unstable.

Typically in power systems, small-signal instability occurs when a complex conjugate pair of eigenvalues moves from having negative real parts (with positive damping ratio) to having positive real parts (with negative damping ratio). Such an instance was encountered in the Western American power system on August 10,

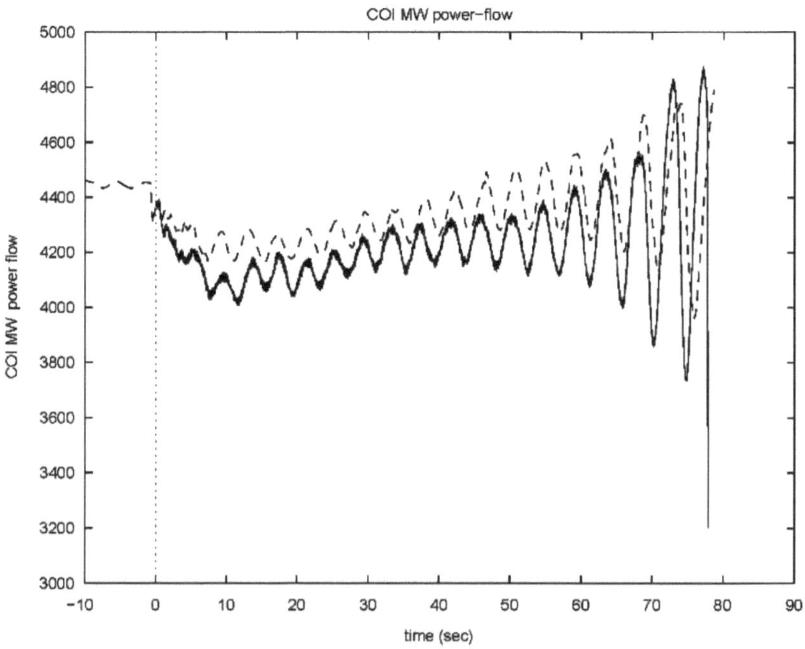

Fig. 3.3 Active power-flow on the California–Oregon tie-line (COI) during the August 10, 1996 blackout event [4]

1996 [4]. Figure 3.3 shows the time-plot of the active MW power-flow on California–Oregon Intertie (COI) 500 kV AC transmission lines on August 10, 1996. The solid curve in Fig. 3.3 shows the actual measured MW flows while the dashed line is the simulated COI MW flow from a validated model that emulated system conditions on that day. Post analysis of the validated model showed that the damping of a 0.25 Hz inter-area oscillatory mode changed from positive damping to negative damping as the event progressed as shown in Fig. 3.4. That is, the complex conjugate eigenvalue pair related to the 0.25 Hz mode moved from the left half complex plane to the right half complex plane during the event, according to the validated model. In Fig. 3.4, Case 0 refers to system conditions just prior to time 0 in the time-plot of Fig. 3.3, while Case 5 corresponds to system conditions at around 60 s in Fig. 3.3. The system conditions about 5 min prior to time 0 in Fig. 3.3 are denoted as Case -1 in Fig. 3.4. Certain 500 kV transmission lines were in service in Case -1, which were later tripped in Case 0 that led to weakening of the damping levels of the 0.25 Hz mode [4]. It is clear from comparison of Figs. 3.3 and 3.4 that the 0.25 Hz oscillations change from a poorly damped response to negatively damped growing oscillations while the eigenvalues move from the left half complex plane into the right half complex plane. From a systems operational view point, it is important to keep all the oscillatory modes of a power system at good positive damping levels so that the system does not encounter either poorly damped or negatively damped oscillations.

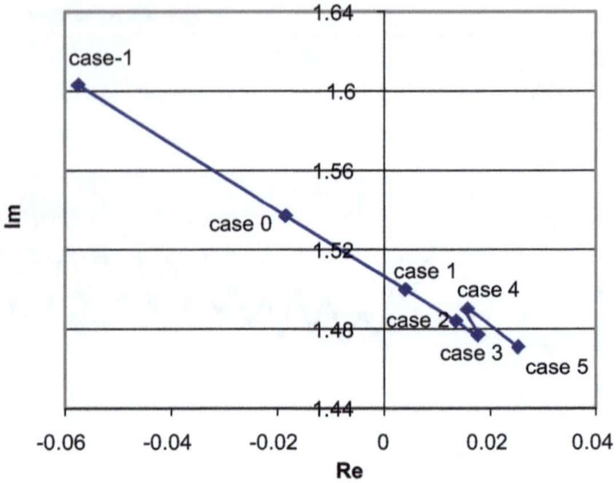

Fig. 3.4 Locus of the 0.25 Hz eigenvalue during August 10, 1996 event from an analysis of the validated model [4]

Poorly damped as well as sustained oscillations can lead to rotor fatigue and can be concerns for power system quality. Negatively damped oscillations or growing oscillations can lead to tripping of generators and can possibly lead to cascading outages such as what happened during August 10, 1996 blackout.

3.1.3 *Transient Stability Phenomena*

Transient stability refers to the system's ability to recover to nominal operating conditions after large disturbances, such as faults, loss of generators, and loss of transmission lines. From a system theory perspective, the system trajectory is no longer near an equilibrium condition and linearization principles do not apply. In the power systems context, the key question is whether the system has enough synchronizing torque or power to keep all the generators in synchronism when subjected to major disturbances. For this reason, the transient stability problem is also referred to as the angle stability problem in some textbooks. A more detailed discussion of transient stability phenomena can be seen in the next chapter and in text books such as [2] and [3].

3.2 Real Time Voltage Stability Monitoring

Power systems can become vulnerable towards voltage instability when any part of the system does not have adequate VAR sources. For example, Sect. 3.1.1 shows that QV plots and QV margins can serve as effective tools in voltage security monitoring. The focus in this section is on real time voltage security monitoring using wide-area synchrophasor measurements. The analysis will be carried out purely using available measurements, and does not require a power-flow model or state estimation model to evaluate the security margins. In other words, the objective of the tools in this section is to estimate the proximity of the system operating condition to static voltage instability limits in terms of the classical QV margin [3] at any bus using synchrophasor measurements. Fundamentally, there are two different approaches for synchrophasor-based voltage stability analysis: (1) Thevenin equivalent-based methods [5, 6] and (2) QV line sensitivity-based methods [7]. In [5] and [6], a Thevenin equivalent reduced order model of the power system is first estimated from real time measurements and the static voltage limit is then calculated on the reduced order model. This method is motivated towards radial type networks that are readily amenable to Thevenin equivalent formulation. The method proposed in [7] estimates QV line sensitivity from synchrophasor measurements using a statistical formulation, and then uses the QV sensitivities for voltage security assessment. This section introduces the QV sensitivity-based method from [7] in more detail with examples.

3.2.1 Voltage Security Index and its Motivation

The example in Sect. 3.1.1 shows that QV plots are effective tools in assessing voltage security of a large power system. If the QV margins are above acceptable thresholds at all the buses in a system, the system can be considered to be voltage secure. However, the QV margins can change rapidly under stressed operating conditions and after major contingencies. Therefore, it is important to keep track of QV margins in real time. As such, QV margins are based on an underlying power-flow model, which may be derived from state estimation. Therefore, the validity of these margins depends on the availability as well as accuracy of the power-flow model. Moreover, computing the QV margins at even a few specified buses is a time-consuming task for real time implementation. The aim of the alternate approach proposed in [7] is to carry out a QV margin-like approach purely using synchrophasor measurements without the need for underlying power-flow models. Basically, it is shown in [7] that the slope of the QV curve at any bus can be estimated from synchrophasor measurements at the bus using statistical methods. And, it can be shown that the slope of the QV curve decreases towards zero as the operation approaches static limits of voltage instability.

Recall that the reactive power injection Q_i at any bus is defined as the sum of all reactive power flows Q_{ij} connected to bus i. That is, $Q_i = \sum_j Q_{ij}$. Therefore, the slope of the QV curve at bus i can be defined as a global voltage security assessment index Γ_i for bus i in the form of:

$$\Gamma_i = \frac{\Delta Q_i}{\Delta V_i} = \sum_j \frac{\Delta Q_{ij}}{\Delta V_i}, \tag{3.1}$$

where ΔQ_{ij} represents reactive power change for each transmission line or transformer (equivalent line model) connected to this bus i. ΔQ_i is an incremental change in bus injection at bus i.

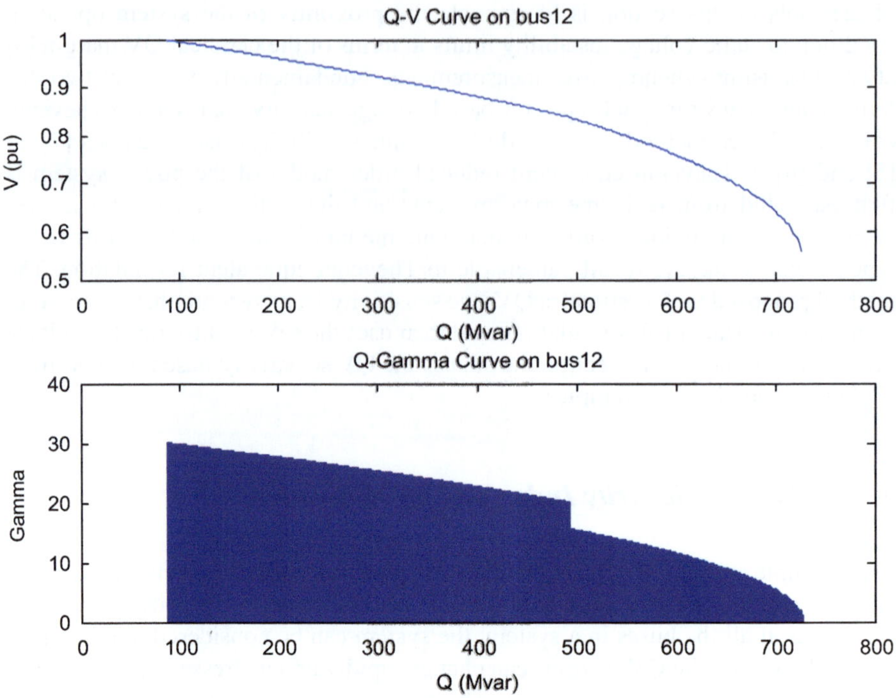

Fig. 3.5 Example of saddle node bifurcation case [7]

It is a well-known fact that the power flow Jacobian becomes singular when the system is at a static saddle-node bifurcation [1]. Hence, the VQ slope at the critical point of QV curve will be infinite. In other words, the security index Γ_i will approach zero in terms of the parameter variation of Q_i at bus i when the variation induces a saddle-node bifurcation at the nose of the QV curve.

In power-flow analysis, there is also the possibility of an abrupt end to the QV curve when the encounter of the reactive power limit (Q limit) at a generator suddenly leads to insolvability of the power-flow equations. This phenomenon has been studied as "Q limit induced bifurcation" [8] or more generally under the name "complementary limit induced bifurcation" in [9]. For the complementary limit induced bifurcation case, the slope $\Gamma_i = \Delta Q_i / \Delta V_i$ will likely stop at some small value instead of approaching zero when the parameter variation induces the limit induced bifurcation.

Figures 3.5 and 3.6 show these two bifurcation cases in the form of QV curves and the corresponding plots of the security index Γ_i. Figure 3.3 is taken as an example for bus 12 on the standard New England 39-bus test system [10]. This is an illustration of the saddle node bifurcation case. At the critical "nose" of QV curve, the index Γ_{12} is very close to 0. Figure 3.6 is an illustration of the complementary limit induced bifurcation case on the IEEE 300-bus test system [10]. For bus 14, the power flow fails when its respective reactive load is approximately 470 MVAr.

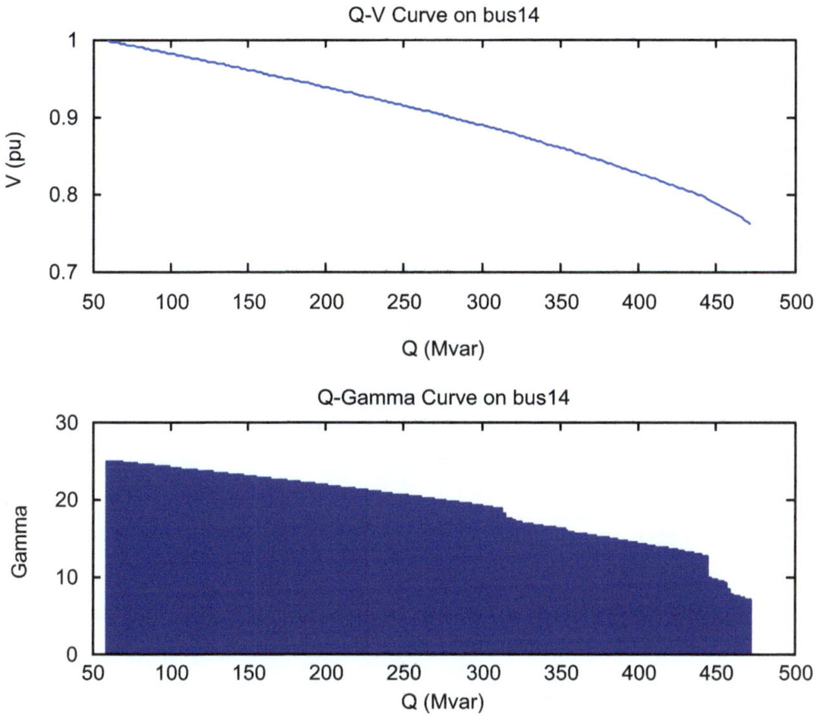

Fig. 3.6 Example of a complementary limit-induced bifurcation case [7]

From the QV curve, the slope around the critical node is not from a smooth "nose" but rather an abrupt edge. This result is reflected well on the "small" value of Γ_{14}, which is 7.01 at the QV limit.

Based on such examples, the following rules are proposed in [7] for using the security index Γ_i towards assessing the system voltage security at any bus *i*.

The index Γ_i can assess bus voltage security through the following approaches:

- A "high" value of Γ_i indicates a "strong" bus in terms of being distant from static voltage instability limit.
- If there is any bus with Γ_i value near 0, the system is close to static voltage stability limit related to saddle-node bifurcation. In this case, the bus with the lowest Γ_i is likely in the critically voltage stressed part of the system.
- If the index Γ_i is less than some critical value, say Γ_i^*, then the system may be vulnerable towards voltage instability caused by either saddle-node or limit induced bifurcations.

In other words, no matter which static bifurcation the system will encounter, the system is operating in a voltage unsecure region, if any bus has a low value for the index Γ_i. In the next section, techniques for computing Γ_i from real-time phasor measurement unit (PMU) measurements are proposed.

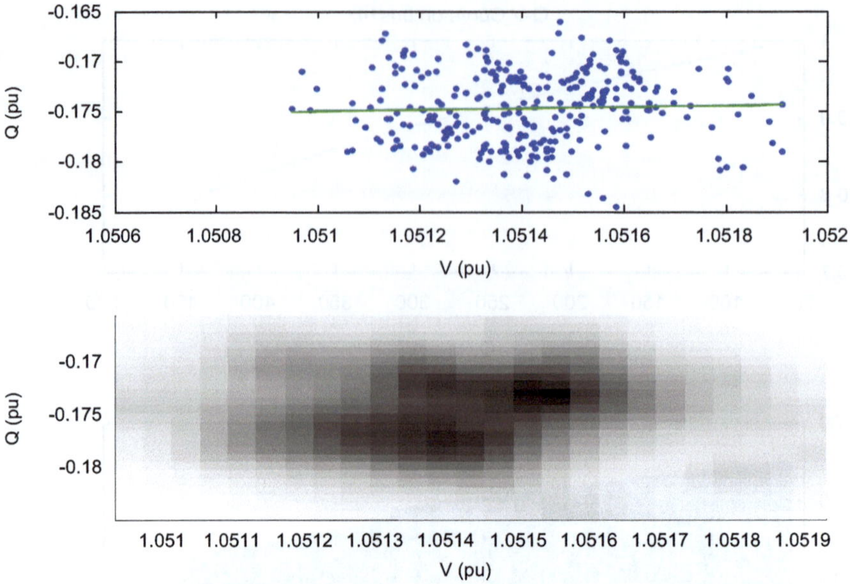

Fig. 3.7 V-Q PMU data points and the density plot for 10ar slope on the QV data; here system PMU [7]

3.2.2 *Estimation of security index Γ_i from PMU data*

In this section, a method for estimating the QV slope Γ_i from real time PMU data is discussed. In this approach, QV line sensitivity on each of the lines connected to bus *i* is estimated first, which leads to an estimation of Γ_i by taking the sum, Eq. (3.1). Accordingly, the problem is to first estimate the QV line sensitivity $\Delta Q_{ij}/\Delta V_i$ from the PMU measurements Q_{ij} and V_i at bus *i*.

First, an attempt is made to extract the slope $\Delta Q_{ij}/\Delta V_i$ for a given transmission line from a sample recorded PMU data set taken from an actual PMU in the Eastern American power system. Figure 3.7 shows a statistical least square fit of a linear slope on the QV data; here the line shows the slope to be near zero. Even the density plot from the data set does not point to any meaningful slope $\Delta Q_{ij}/\Delta V_i$ for this line.

However, based on a capacitor bank-switching event near the bus, it could be concluded that the slope $\Delta Q_{ij}/\Delta V_i$ was definitely not zero. The apparent inconsistency as to why the statistical fit failed to reveal QV sensitivity could be explained by causality of the QV relationship. In the real system, note that there are actually two slopes $\Delta Q_{ij}/\Delta V_i$ related to the transmission line from bus *i* to bus *j* as shown in Fig. 3.8. Whether the change in line-flow is caused by voltage changes at the sending end of the line or the receiving end of the line will result in two different slopes with opposite signs. Therefore, the QV slope should be estimated only by first dividing the data into two different data sets, each with a different causality as shown next.

When the same data in Fig. 3.7 are split according to positive slopes and negative slopes as shown in Fig. 3.8, the trend of the data points lining up along a definite

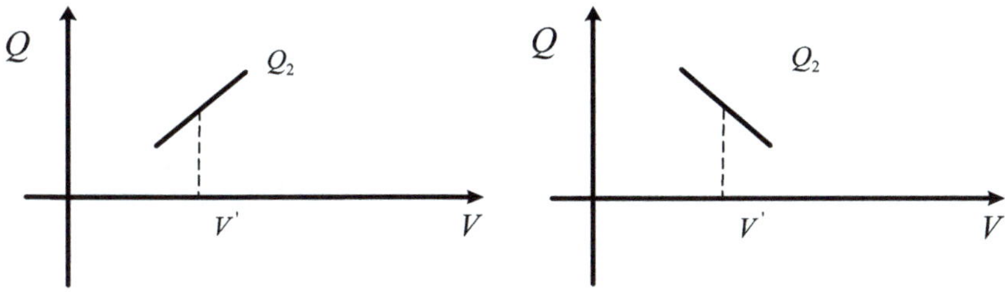

Fig. 3.8 Two different VQ relationships between reactive power Q_{ij} and bus voltage V_i in any small time interval [7]

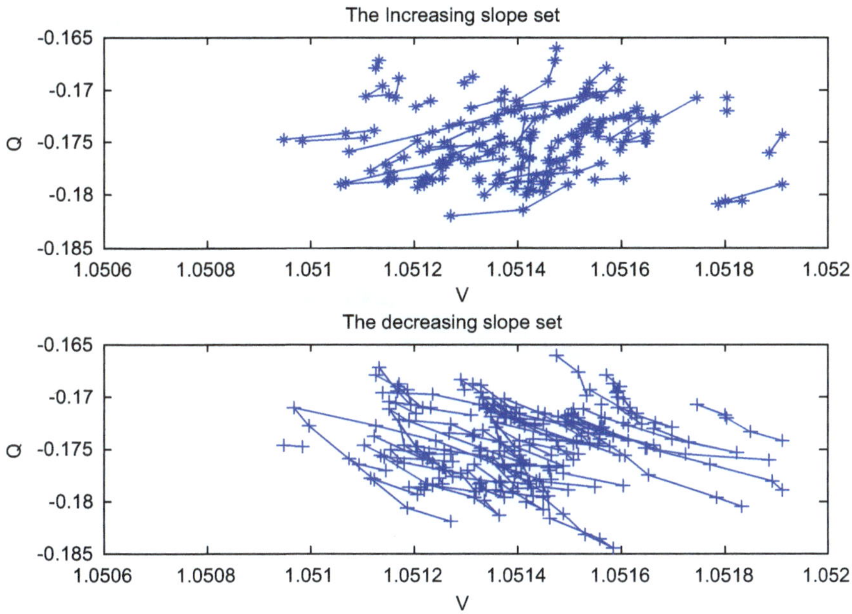

Fig. 3.9 Two sets of PMU V versus Q points with increasing and decreasing slopes after data split [7]

positive slope versus a clearly defined negative slope for the QV line sensitivity can clearly be seen. Using statistical analysis, the two slopes can then be estimated. The slopes from these te bient" PMU responses also match very well with sensitivities calculated from discrete switching events (Fig. 3.9).

3.2.3 Estimation Example Based on Actual PMU Data

This is a test set of PMU data where the reactive power line-flow shows a jump around 35very well with Fig. 3.10. Figure 3.11 shows the PMU data subsets after

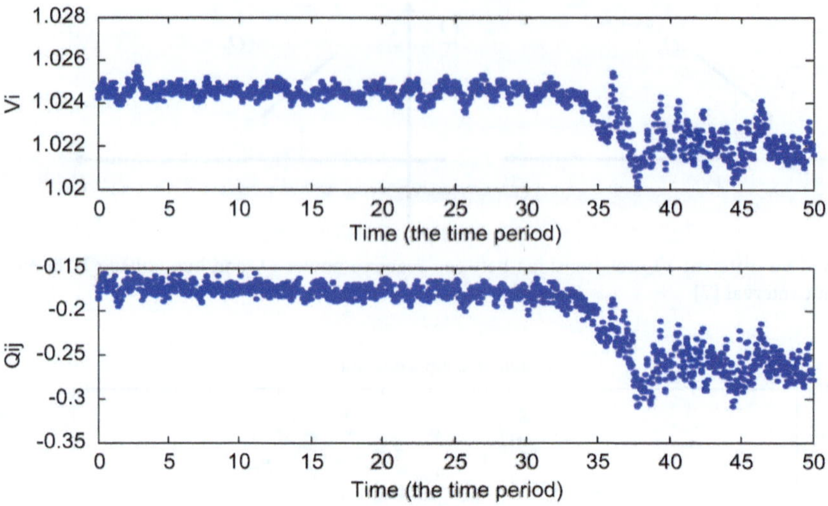

Fig. 3.10. Time-plots of V_i and Q_{ij} for the time period from 0 to 50ated [7]

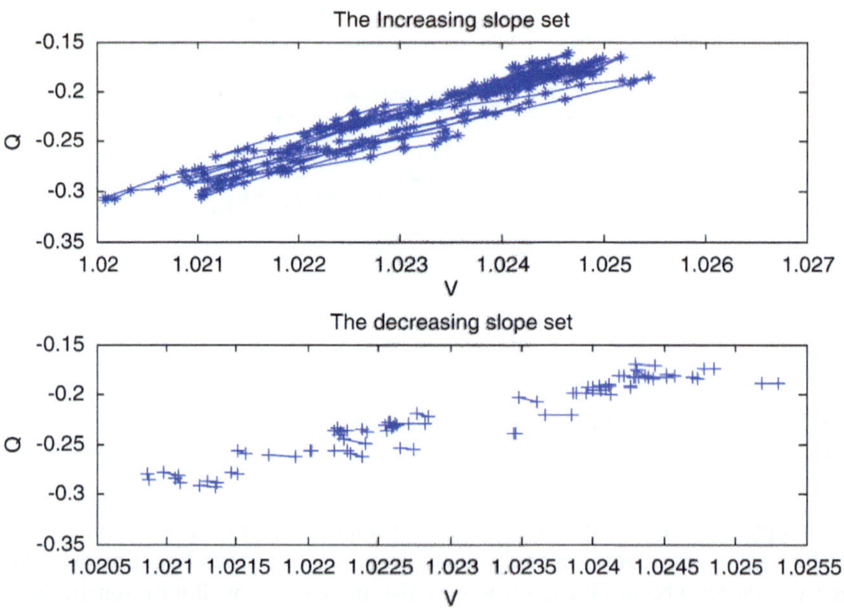

Fig. 3.11 PMU VQ data pairs for the time period from 30 to 40js after data split [7]

a data split for the time period between 30 and 40d 35 Table 3.1 summarizes the results of the statistical data analysis. In Table 3.1, the sensitivity α denotes the estimated positive slope while the sensitivity β denotes the corresponding negative slope for the QV sensitivity $\Delta Q_{ij}/\Delta V_i$ for this line. The value α_{Jump} is the positive slope for $\Delta Q_{ij}/\Delta V_i$ calculated directly from the discrete event near time 35itivity d

Table 3.1 QV line sensitivities from PMU data after data split [7]

Sensitivity	0–1 s	0c10 s	30 31 s	30 40 s	40 41 s	40 50 s
A	38.7	38.4	49.5	41.3	28.9	36.1
B	−22.4	−28.4	−42.0	−30.4	−27.9	−40.7
α_{jump}	42.3 around 35 s					

between 30 and 40d 35very well with sensitivities calculated from discrete switchα and β in Table 3.1. It can be seen that the estimated slopes in Table 3.1 match well with the slope calculated from discrete jump, especially for the longer span of PMU data over 10-s intervals.

Based on a similar analysis, the line sensitivities $\Delta Q_{ij}/\Delta V_i$ can be calculated for each of the lines connected to bus i giving us the net sum Γ_i in (3.1), which is then the voltage stability analysis (VSA) index for bus i. Whenever the value Γ_i goes below a predefined threshold, say Γ_i^*, the system near the monitored bus approaches a static voltage stability limit, as discussed earlier in Sect.ivities

3.2.4 Real Time Implementation from PMU Data

A flow chart implementing the QV sensitivity-based security assessment algorithm discussed in Sects.voltage stability limit, as discuFig. 3.12.

In the flow chart in Fig. 3.12, it is assumed that the PMU data from available PMUs is first time-aligned by the phasor data concentrator (PDC). Next, the algorithm computes the QV line sensitivity for each line monitored by a PMU after doing the data split, as discussed in Sect. 3.2.3. It then computes the overall security index Γ_i for the bus by the sum of individual line sensitivities from Eq. (3.1). Next, vulnerability to voltage insecurity is assessed by comparing whether each index Γ_i is below its threshold of Γ_i^*. Relative values of Γ_i with respect to Γ_i^* among different buses then provide an indication of the weakest part of the system for potential control action and for alerting or alarming the system operators. The complete analysis is then repeated every few seconds or minutes as configured by the system administrator.

3.3 Real Time Oscillation Monitoring

Worldwide large-scale implementations of synchrophasors in recent years have motivated significant research on measurement-based oscillation monitoring algorithms in the power engineering literature [11]. Such measurement-based algorithms can be broadly divided into two categories: (1) ringdown analysis algorithms and (2) ambient data algorithms. In ringdown analysis methods, the power system response following medium-scale disturbances such as loss of a transmission line or a generator is analyzed using Prony methods for extracting the dominant modal

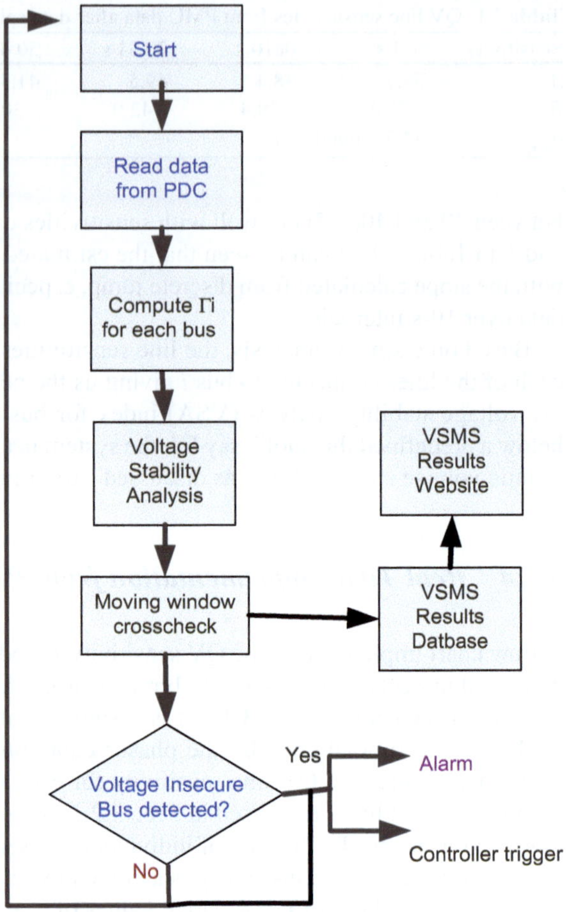

Fig. 3.12 Flow chart for voltage security assessment using wide-area *PMU* data. *PDC* phasor data concentrator, *VSMS* - Voltage Stability Monitoring System.

content in the PMU data [12]. Therefore, ringdown analysis methods are only useful when the system response following a disturbance or other oscillatory events is available. On the other hand, ambient data methods are applicable for power system oscillation monitoring at all times. This is because ambient methods analyze power system response to small random fluctuations always present in the loads that are usually modeled by white noise input. Ambient methods are useful for extracting the dominant modal properties of oscillatory modes by analyzing power system responses to white noise-type load perturbations. In this section, some of the major algorithms used in PMU-based oscillation monitoring are introduced.

Figure 3.13 illustrates two kinds of PMU data in oscillation analysis. The system undergoes a transmission line tripping at around 835 s in Fig. 3.13. The system experiences a ringdown response between 835 and 850 s, and the large number of swings during the ringdown indicates the presence of a poorly damped oscillatory mode. By fitting a sum of exponentially decaying oscillatory modal responses, Prony analysis of the ringdown shows the oscillatory mode to be at about 1.2 Hz

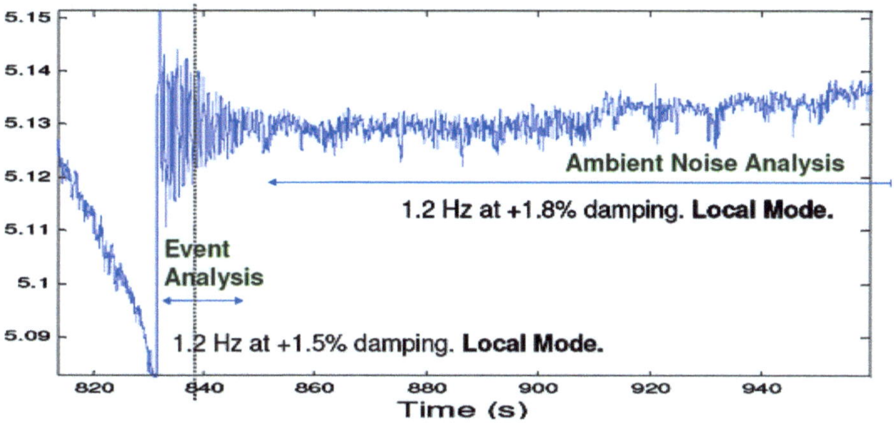

Fig. 3.13 Two kinds of data in oscillation monitoring using PMUs [13]

with approximately 1.5 % damping ratio [13]. After 850 s, the system is responding to random load fluctuations as indicated by the noise-like response of the PMU measurement in Fig. 3.13. Signal processing algorithms can be applied to this ambient response, and details will be presented in Sect. 3.3.3. It can be shown [13] that the ambient methods provide similar estimation results of the poorly damped, comparable to Prony methods as indicated in Fig. 3.13.

3.3.1 Oscillation Monitoring System

An oscillation monitoring system (OMS) has been proposed in [13] for wide-area monitoring of power system oscillations from real time PMU measurements. A flow chart of the OMS is presented in Fig. 3.14. The OMS combines Prony methods and ambient data methods for oscillation analysis into a unified framework. Similar to the case of real time voltage security monitoring in Sect.onentially decaying oscipplicable for power system oscillation monitoring at all detection algorithms are applied. Whenever an event is detected, the engine treats the subsequent data, say over the following 20 s (prespecified time interval), as ringdown response and carries out ringdown analysis methods under the umbrella of ella of nalysis engine" [13]. Normal ambient data, on the other hand, is processed using ambient algorithms under the framework of lla of oltage security monito13]. In each case, raw estimates from signal processing algorithms are processed via consistency checks to ensure the linearity of system responses towards their validity in small-signal analysis of power system measurements. This is important because power systems typically encounter nonlinear characteristics, such as saturation and hard limits in the time periods immediately following disturbances. Moreover, discrete switching events such as insertion and tripping of shunt and series capacitor banks also need to be

Fig. 3.14 Flowchart of an oscillation monitoring system [13]. *PDC* phasor data concentrator, *FDD* frequency domain decomposition

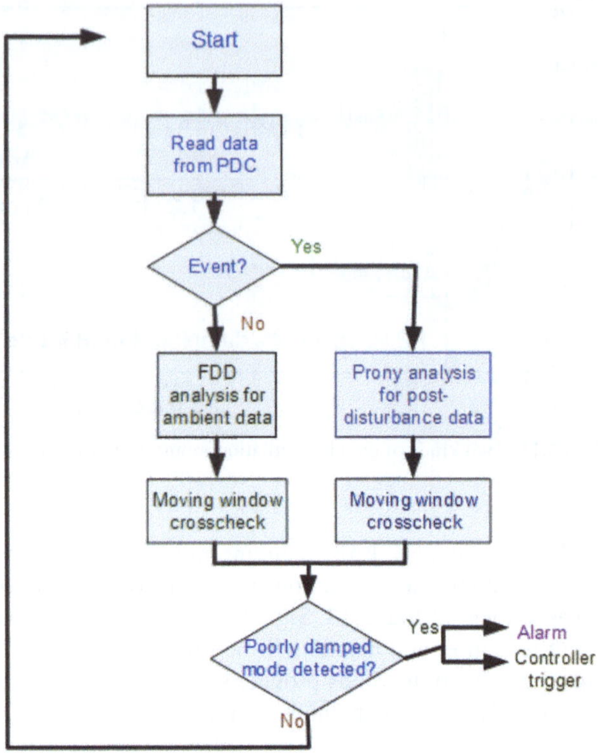

handled with caution in automatic real time analysis of PMU data. It is important to keep in mind the limitations of the analysis methods so that the estimation results are consistent with the assumptions made in deriving the estimation methodologies. Prony algorithms and event analysis engine are discussed in more detail in Sect. Sect.s immediatelethods and damping monitor engine are introduced with examples in Sect. 3.3.3.

3.3.2 Event Analysis Engine and Prony Algorithms

The event analysis engine provides two levels of oscillation detection using real time analysis of wide-area measurements following a power system disturbance. That is, the OMS is able to detect local modes as well as inter-area modes. For the local oscillation detection, multiple signals from the same PMU (or multiple PMUs located at the same substation) are used in Prony analysis. The signal groups used for local analysis are prespecified. However, the inter-area mode signal groups are selected automatically from those PMUs that participate in the specific inter-area oscillatory mode. All these tasks of local oscillation detection can be executed in parallel by multithreading in powerful servers exclusively for oscillation monitoring in the control center. The approach is inherently multidimensional and aimed at analyzing hundreds of PMU measurements in real time automatically.

After the local mode analysis, the next step is to determine the dominant mode in each PMU for each method, and then the crosscheck for all methods is performed in each PMU. The results are considered to be consistent if the dominant modes from different methods fall in a specified frequency range and damping ratio range. The above crosscheck is called local mode crosscheck. If more than two PMUs have consistent local crosschecks, then the inter-area mode detection tasks are activated. Inter-area mode crosscheck is performed from the subset of PMUs showing oscillations of the same frequency. Moving window crosscheck is utilized to compare results along the sliding data windows. If several moving window crosschecks give consistent estimates, e.g., four consecutive consistent moving windows crosschecks, then the resulting damping ratio estimates can be classified to be reliable. This rule is motivated by the underlying assumption of linearity of the ringdown responses. For a purely linear system response, the damping estimates will be the same across multiple moving time windows whereas the estimates typically change when applied to responses from nonlinear systems. Reliable estimates that pass the consistency checks can be compared to a threshold in order to determine whether an alert needs to be triggered. For instance, the damping ratio threshold for detecting poorly damped oscillations could be set to be say +3 % for reliable operation of power systems. The theoretical background for the event monitor engine is repeated from [14] first, before examples are presented later in the section.

The power system is basically a high-order nonlinear system. However, for small disturbances, the system can be linearized around its operation point (or equivalent point)

$$\begin{aligned} \Delta\dot{x} &= A\Delta x + b\Delta u \\ \Delta y_i &= c_i \Delta x \end{aligned} \quad , i = 1,2,\ldots,m,$$

where Δx is the linearized state vector, A is the linearized system matrix, b and c are the input and output state matrices, respectively, Δu is the input, and Δ_{yi} is the i-th output. As shown in [3], the transfer functions between the input and output have the following form

$$G_i(s) = \frac{\Delta y_i(s)}{\Delta u(s)} = \sum_{i=1}^{n} \frac{R_i}{s - \lambda_i}, \tag{3.2}$$

where $R_i = c_i \varphi_i \psi_i b$ and φ_i and ψ_i are right eigenvector and left eigenvector corresponding to λ_i, respectively.

If an impulse is applied as input to the system, the m outputs are

$$y_j(t) = \sum_{i=1}^{n} R_i \, exp(\lambda_i t), \; j = 1,2,\ldots m. \tag{3.3}$$

If the input is not an impulse, one can get a similar form of the outputs. That is, a sum of exponential terms can be derived. This is the form that the proposed methods can be applied to for parameter estimation. When $y_j(t)$ is sampled at a constant sampling period Δ_t, we get the following discrete form.

$$y(k) = \sum_{i=1}^{n} R_i z_i^k, \tag{3.4}$$

where $z_i = exp(\lambda_i \Delta t)$, $\lambda_i = \sigma_i + j\varpi_i \cdot n$ is called the model order, which is not known for real power system measurements.

Pronyis method First, Eq. (3.4) can be written in the following form:

$$\begin{bmatrix} y(0) \\ y(1) \\ \\ y(N-1) \end{bmatrix} = \begin{bmatrix} 1 & 1 & \cdots & 1 \\ z_1 & z_2 & \cdots & z_n \\ \cdots & \cdots & \cdots & \cdots \\ z_1^{N-1} & z_2^{N-1} & \cdots & z_n^{N-1} \end{bmatrix} \begin{bmatrix} R_1 \\ R_2 \\ \cdots \\ R_n \end{bmatrix}. \tag{3.5}$$

The z_i's are necessarily the roots of an nth-order polynomial with unknown coefficients a_i, and they satisfy

$$z^n - (a_1 z^{n-1} + a_2 z^{n-2} + \ldots + a_n z^0) = 0. \tag{3.6}$$

Left-multiply $\left[-a_n, -a_{n-1}, \ldots, -a_1, 1, 0, \ldots 0\right]$ to the both sides of Eq. (3.5) and by using (3.6), the following equation is obtained:

$$\left[-a_n, -a_{n-1}, \ldots, -a_1, 1, 0, \ldots 0\right] \begin{bmatrix} y(0) \\ y(1) \\ \\ y(N-1) \end{bmatrix} = 0. \tag{3.7}$$

Left-multiply $\left[0, -a_n, -a_{n-1}, \ldots, -a_1, 1, 0, \ldots 0\right]$ to both sides of Eq. (3.5) and the resulting right-hand side is also zero. Equation (3.7) can be repeated to get the following form:

$$\begin{bmatrix} y(n) \\ y(n+1) \\ \\ y(N-1) \end{bmatrix} = \begin{bmatrix} y(n-1) & y(n-2) & \cdots & y(0) \\ y(n) & y(n-1) & \cdots & y(1) \\ \cdots & \cdots & \cdots & \cdots \\ y(N-2) & y(N-3) & \cdots & y(N-n-1) \end{bmatrix} \begin{bmatrix} a_1 \\ a_2 \\ \cdots \\ a_n \end{bmatrix}. \tag{3.8}$$

Now, the procedure of Prony's method can be summarized in the following three steps:

Step 1. Solve Eq. (3.8) to get the coefficients a_i.

Step 2. Calculate the roots of Eq. (3.6) to get z_i.

Step 3. Solve Eq. (3.5) for complex residues R_i.

Prony's method can be extended to analyze multiple signals simultaneously. In step 1, the equations for each signal in Eq. (3.8) are stacked. Consider a set of m signals $y_j(t)$ with $j=1, 2,\ldots,m$. Now a total of $(N-n)\times m$ equations with n unknown coefficients a_i in Eq. (3.9) are obtained. The coefficients a_i are solved for in the least-square sense. After calculating roots z_i, step 3 results in the solution of the following equations:

$$\begin{bmatrix} y_1(0) & y_2(0) & \cdots & y_m(0) \\ y_1(1) & y_2(1) & \cdots & y_m(1) \\ \cdots & \cdots & \cdots & \cdots \\ y_1(N-1) & y_2(N-1) & \cdots & y_m(N-1) \end{bmatrix}$$

$$= \begin{bmatrix} 1 & 1 & \cdots & 1 \\ z_1 & z_2 & \cdots & z_n \\ \cdots & \cdots & \cdots & \cdots \\ z_1^{N-1} & z_2^{N-1} & \cdots & z_n^{N-1} \end{bmatrix} \begin{bmatrix} R_1^1 & R_1^2 & \cdots & R_1^m \\ R_2^1 & R_2^2 & \cdots & R_2^m \\ \cdots & \cdots & \cdots & \cdots \\ R_n^1 & R_n^2 & \cdots & R_n^m \end{bmatrix}. \tag{3.9}$$

A practical issue in Prony's method is to determine the unknown model order n. The common procedure is to fit a high order model to the data, and the modes corresponding to the noise having small residue magnitudes can be filtered out from the result.

Matrix Pencil Method First two matrices Y_1 and Y_2 are defined as shown below:

$$Y_1 = \begin{bmatrix} x(0) & x(1) & \cdots & x(L-1) \\ x(1) & x(2) & \cdots & x(L) \\ \cdots & \cdots & \cdots & \cdots \\ x(N-L-1) & x(N-L) & \cdots & x(N-2) \end{bmatrix}, \tag{3.10}$$

$$Y_2 = \begin{bmatrix} x(1) & x(2) & \ldots & x(L) \\ x(2) & x(3) & \ldots & x(L+1) \\ \ldots & \ldots & \ldots & \ldots \\ x(N-L) & x(N-L+1) & \ldots & x(N-1) \end{bmatrix}, \tag{3.11}$$

where $x(k)$ are the noise-free data points. L is the pencil parameter. Using Eq. (3.4), Eqs. (3.10) and (3.11) can be written as

$$Y_2 = Z_1 R Z_0 Z_2, \tag{3.12}$$

$$Y_1 = Z_1 R Z_2, \tag{3.13}$$

where

$$Z_0 = \text{diag}\left[z_1, z_2, ..., z_n\right],$$

$$Z_1 = \begin{bmatrix} 1 & 1 & \dots & 1 \\ z_1 & z_2 & \dots & z_n \\ \dots & \dots & \dots & \dots \\ z_1^{N-L-1} & z_2^{N-L-1} & \dots & z_n^{N-L-1} \end{bmatrix}_{(N-L)\times n},$$

$$Z_2 = \begin{bmatrix} 1 & z_1 & \dots & z_1^{L-1} \\ 1 & z_2 & \dots & z_2^{L-1} \\ \dots & \dots & \dots & \dots \\ 1 & z_n & \dots & z_n^{L-1} \end{bmatrix}_{n\times L},$$

$$R = \text{diag}\left[R_1, R_2, ...R_n\right].$$

Here diag[z_1,…,z_n] represents a diagonal matrix consisting of the elements z_1 through z_n as its diagonal entries. The matrix pencil is defined as follows:

$$Y_2 - \lambda Y_1 = Z_1 R\left\{Z_0 - \lambda I\right\} Z_2, \tag{3.14}$$

where I is the identity matrix.

When $n \le L \le N - n$, the rank of the matrix $Y_2 - \lambda Y_1$ is n if $\lambda \neq z_i$. However, if $\lambda = z_i$, then the ith row of the matrix $Z_0 - \lambda I$ becomes zero, and the rank of $Y_2 - \lambda Y_1$ is reduced by one. Hence, the parameters z_i's are the generalized eigenvalues of the matrix pair $\left\{Y_2; Y_1\right\}$. Or, equivalently, one can solve the ordinary eigenvalues of the matrix $Y_1^+ Y_2 - \lambda I$ to get the parameters z_i, where Y_1^+ is pseudo-inverse of Y_1. After solving for parameters z_i, we solve Eq. (3.5) for residues.

For the actual measured data, define a new matrix Y containing the noisy data as follows:

$$Y = \begin{bmatrix} y(0) & y(1) & \dots & y(L) \\ y(1) & y(2) & \dots & y(L+1) \\ \dots & \dots & \dots & \dots \\ y(N-L-1) & y(N-L) & \dots & y(N-1) \end{bmatrix}. \tag{3.15}$$

When comparing Eqs. (3.10) and (3.11), it can be seen that Y_1 is obtained by deleting the last column of Y, and Y_2 is obtained by deleting the first column of Y. Next, apply Singular Value Decomposition (SVD) to Y as follows:

$$Y = U\Sigma V^*, \tag{3.16}$$

where U and V are unitary matrices. Σ is a diagonal matrix containing the singular values of Y with descending order. The superscript * denotes conjugate transpose.

If the data are noise free, the matrix Y has n nonzero singular values. However, when noise is present, the zero singular values are perturbed and become nonzero. Now, the singular value decomposition provides an effective way of noise filtering. In the singular values below, some specified thresholds are considered to be caused by noise and need to be reset to zero values.

Then, the n significant singular values are used to reconstruct the original data matrix as shown below.

$$V' = [v_1, v_2, \ldots, v_n], \tag{3.17}$$

$$Y_1 = U\Sigma' V_1'^*, \tag{3.18}$$

$$Y_2 = U\Sigma' V_2'^*, \tag{3.19}$$

where $v_i's$ are column vectors of V corresponding to the n dominant singular values. Σ' is the first n columns of Σ. V_1' is obtained from V' with the last row of V' deleted. V_2' is obtained from V' with the first row of V' deleted. This yields

$$Y_1^+ Y_2 = (V_1'^* V_1'^*)^{-1} V_1' \times V_2'^* = (V_1'^*)^+ V_2'^*.$$

The nonzero eigenvalues of $(V_1'^*)^+ V_2'^*$ are the same as the eigenvalues of $V_2'^* (V_1'^*)^+$, where $V_2'^* (V_1'^*)^+$ is an $n \times n$ matrix. Calculating the eigenvalues of an $n \times n$ matrix is computationally inexpensive since n is usually a small number. After the eigenvalues are obtained, solve Eq. (3.5) for the residues. The computationally expensive step in the matrix pencil algorithm is the computation of SVD in Eq. (3.16).

Hankel Total Least Square Method First, form the Hankel matrix as follows:

$$H = \begin{bmatrix} y_0 & y_1 & \cdots & y_{M-1} \\ y_1 & y_2 & \cdots & y_M \\ \cdots & \cdots & \cdots & \cdots \\ y_{L-1} & y_L & \cdots & y_N \end{bmatrix}, \tag{3.20}$$

where L is a parameter chosen to be larger than n, $Ma=N+1-L$, and N is the number of measurements. If there were no noise, i.e., $y(k) = \sum_{i=1}^{n} R_i z_i^k$, then H can be factorized as follows:

$$H = \begin{bmatrix} 1 & \cdots & 1 \\ z_1^1 & \cdots & z_n^1 \\ \cdots & \cdots & \cdots \\ z_1^L & \cdots & z_n^L \end{bmatrix} \begin{bmatrix} R_1 & & \\ & \ddots & \\ & & \ddots & \\ & & & R_n \end{bmatrix} \begin{bmatrix} 1 & z_1^1 & \cdots & z_1^{M-1} \\ \cdots & \cdots & \cdots & \cdots \\ \cdots & \cdots & \cdots & \cdots \\ 1 & z_n^1 & \cdots & z_n^{M-1} \end{bmatrix} = SRT^T, \tag{3.21}$$

where both S and T are Vandermonde matrices. The matrix S is shift-invariant, that is

$$S_{\downarrow} Z = S_{\uparrow}, \tag{3.22}$$

where the up (down) arrow placed behind a matrix stands for deleting the top (bottom) row of the matrix and Z is the $n \times n$ diagonal matrix whose diagonal entries are, z_1,di z_n.

The Hankel matrix can also be decomposed by SVD as follows:

$$H = U \Sigma V^* = \begin{bmatrix} \hat{U} & U_0 \end{bmatrix} \begin{bmatrix} \hat{\Sigma} & \\ & \Sigma_0 \end{bmatrix} \begin{bmatrix} \hat{V} \\ V_0 \end{bmatrix}^*, \tag{3.23}$$

where U and V are unitary matrix, i.e., $UU^* = I$, $VV^* = I$ and I is the identity matrix. Σ is a diagonal matrix with singular values on the diagonal in decreasing order. $\hat{\Sigma}$ is the submatrix containing first n singular values. If there is no noise in the signal, the submatrix Σ_0 is null. Otherwise, Σ_0 is a full matrix with small singular values along the diagonal. When noise is present, as in the case of PMU measurements, the usual procedure is to set a prespecified threshold for noise-related singular values. Those singular values below the threshold are considered to be generated by noise and can be discarded.

In the noise-free case, $H = \hat{U} \hat{\Sigma} \hat{V}^*$. Comparing this equation with Eq. (3.21), the following relationship is obtained:

$$\hat{U} = SQ, \tag{3.24}$$

where Q is an $n \times n$ nonsingular matrix. Deleting the first row in the above equation gives $\hat{U}_{\downarrow} = S_{\downarrow} Q$, while deleting the last row gives $\hat{U}_{\uparrow} = S_{\uparrow} Q$. Therefore, $\hat{U}_{\downarrow}$ and $\hat{U}_{\uparrow}$ are related by the following equation:

$$\hat{U}_{\uparrow} = S_{\downarrow} Z Q = \hat{U}_{\downarrow} Q^{-1} Z Q = \hat{U}_{\downarrow} \tilde{Z}, \tag{3.25}$$

where $\tilde{Z} = Q^{-1} Z Q$, which has the same eigenvalues as Z, i.e., z_1,i. z_n.

In the noisy case, Eq. (3.25) does not hold exactly. In this case, $\tilde{Z}$ can be solved by a total least square method. After calculating $\tilde{Z}$, the signal poles are computed as the eigenvalues of $\tilde{Z}$.

When multiple signals need to be analyzed simultaneously, simply replace H in Eq. (3.20) with the horizontally stacked Hankel matrices as follows:

$$H_{\text{series}} = [H_1 H_2 \cdots H_m], \tag{3.26}$$

where m is the number of signals. The last step to calculate residues R_i' is the number of signals. The laPronyhe method.

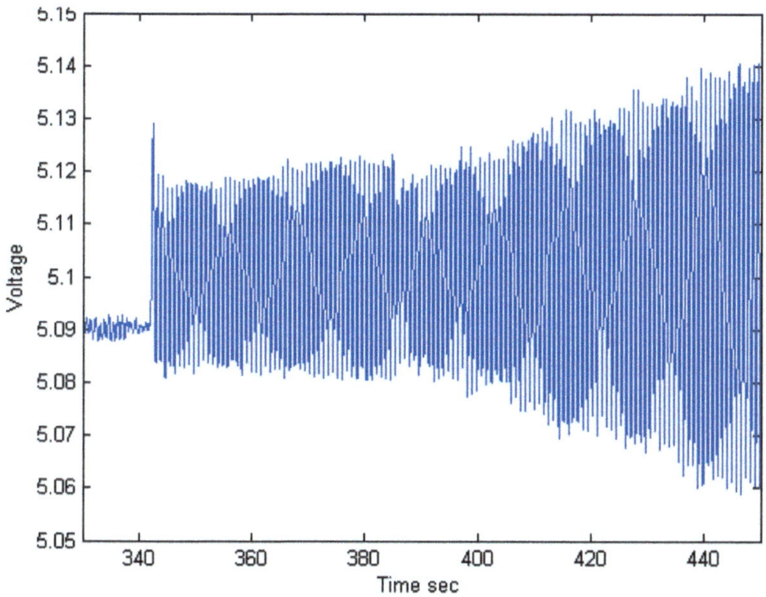

Fig. 3.15 Bus voltage magnitude at a 500 kV bus during an eastern interconnection event [15]

Example The following example is taken from a recent event in the Eastern American interconnection. The disturbance resulted in oscillations shown in Fig. 3.15 from a local mode, which was primarily seen in one PMU (denoted PMU6 in this example). Signals from each PMU are analyzed individually first as part of local PMU analysis. Accordingly, OMS issues a local estimate when moving window cross-check reaches a consistent estimate. Then, signals from all PMUs that are reporting consistent estimates within a common frequency range are grouped together and inter-area analysis is initiated. When consistent estimates are seen from multiple PMU groups, inter-area estimates are issued, and depending on the observed damping levels, operator alarms can be issued.

The voltage magnitude at PMU6 for the event is shown in Fig. 3.15. The event monitor engine analyzes multiple signals from eight PMUs and issues a consistent damping estimate of 0.09 % damping of a 1.18 Hz mode at 352 s in the plot.

Figure 3.16 shows a summary of local analysis results from the eight PMUs. It can be seen that PMU6 shows the most consistent estimates whereas the mode is also seen less so in the other PMUs. An example of the local analysis results is shown for PMU6 in Figs. 3.17 and 3.18 for the mode frequency estimates and the mode damping estimates, respectively. It can be seen that the estimates from the three engines, namely, Prony, matrix pencil, and Hankel total least square (HTLS) become consistent after about 345 s in Figs. 3.17 and 3.18. The estimates are also consistent across multiple moving window calculations, which confirms the analysis to be linear towards the validity of Prony type of analysis. Conversely, the

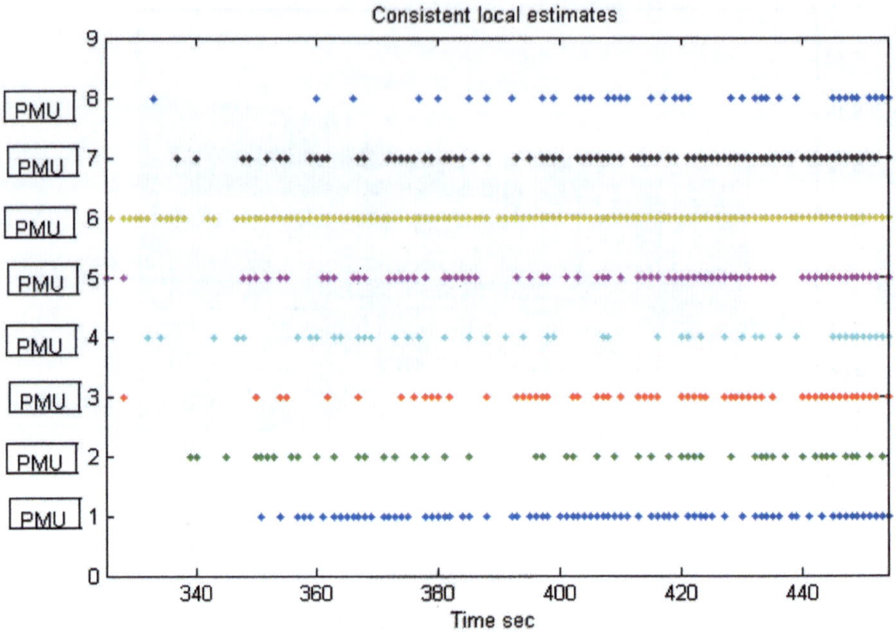

Fig. 3.16 Summary of local analysis results for event analysis engine [15]. *PMU* phasor measurement unit

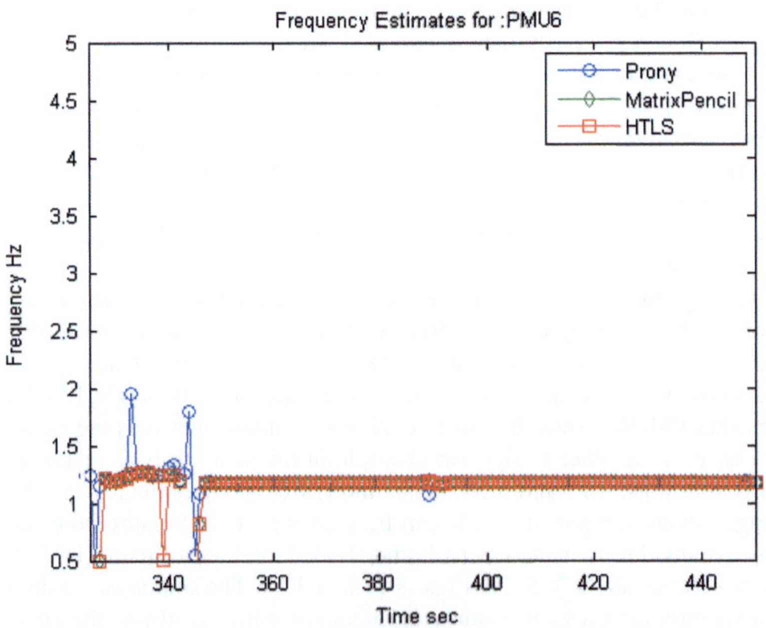

Fig. 3.17 Plot of mode frequency from local analysis estimates for PMU6 [15]. *PMU* phasor measurement unit, *HTLS* Hankel total least square

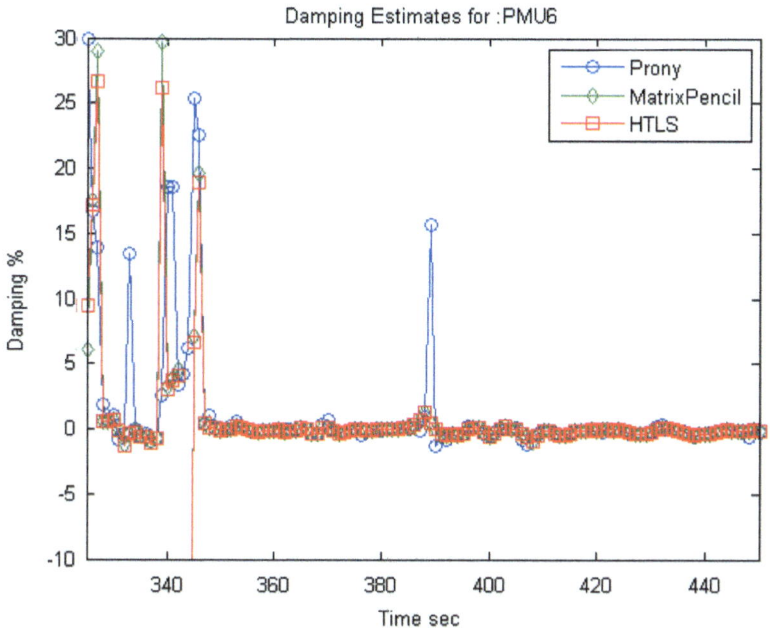

Fig. 3.18 Plot of mode damping ratio from local analysis estimates for PMU6 [15]. *PMU* phasor measurement unit, *HTLS* Hankel total least square

estimates change considerably during the time period prior to 345 s indicating the potential presence of nonlinear responses and/or switching events in the Prony time window. Accordingly, such inconsistent estimates are dropped from analysis and they are not reported out of the event analysis engine.

Next, estimation results from inter-area analysis of multiple PMUs 1, 5, 6, and 7 that are chosen by the engine for inter-area analysis are shown in Figs. 3.19 (mode frequency) and 3.20 (mode damping ratio). Consistency checks are applied across multiple engine estimates as well as across multiple consecutive moving window estimates as in the case of the local event analysis engines. From Figs. 3.19 and 3.20, it can be seen that the estimates are not consistent during the initial phase of the disturbance prior to about 350 s and the estimates do become consistent just after 350 s.

The time-plot of the bus voltage in Fig. 3.15 shows the presence of an unknown switching event at about 385 s, thereby changing the damping level of the mode. A review of the corresponding estimates of the local and inter-area engine estimates shows that the results are inconsistent just after the disturbance, whereas they settle down to consistent values shortly after the disturbance. The proposed methodology serves as an automatic framework for screening nonlinear portions of power system PMU responses from adversely affecting the linear modal estimates of Prony type engines in the event analysis engine. Basic principles of linearity are exploited in the form of consistency checks that serve to eliminate nonlinearity induced unreliable estimates from Prony type curve fitting-based analysis.

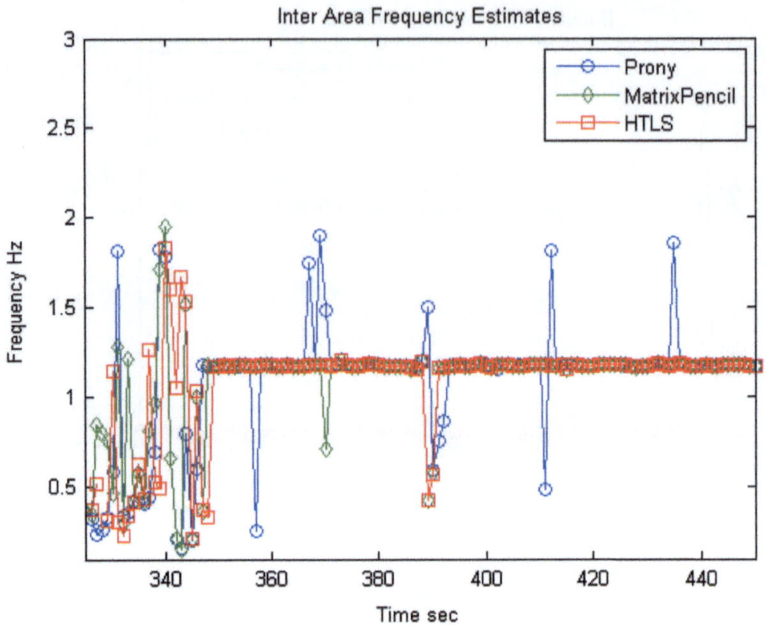

Fig. 3.19 Plot of mode frequency from inter-area analysis estimates of multiple PMUs [15]. *HTLS* Hankel total least square

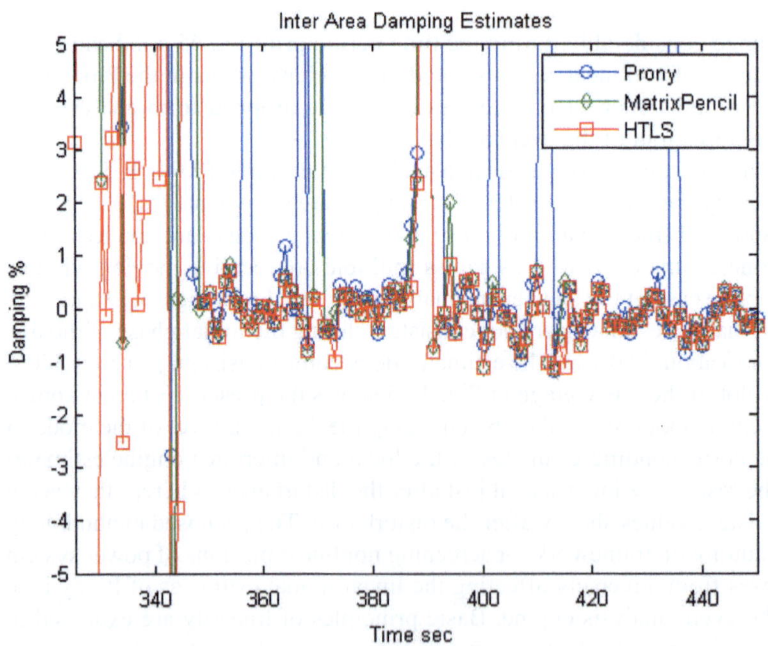

Fig. 3.20 Plot of mode frequency from inter-area analysis estimates of multiple PMUs [15]. *HTLS* Hankel total least square

3.3.3 Damping Monitor Engine and Ambient Algorithms

Low-frequency electromechanical oscillations have remained a major concern for long-distance power transfers in many power systems. Insufficient damping of power systems can lead to oscillations with growing magnitudes as in the case of the Western American blackout on August 10, 1996, discussed in Sect. 3.1.2. The problem of detecting and controlling these low-frequency electromechanical oscillations has been a challenging research topic for many years. For oscillation monitoring, the ambient (or routine) measurement type methods for oscillation monitoring are attractive. All ambient-type methods are applied to ambient measurements of power systems, i.e., the measurements when the system is in a normal operating condition are subject to routine random load fluctuations. The main advantage of ambient data is that these methods work in a nonintrusive manner, and they are always available. On the other hand, modal estimates from the Prony algorithms discussed in Sect. 3.3.2 are only available for a short period after system disturbances.

The first study on ambient data in power systems [16] models the measurements as the output from an autoregressive (AR) model driven by white noise. There are several other algorithms based on techniques, such as the autoregressive moving average (ARMA) methods and the Stochastic Subspace Identification (SSI) method, details of which can be seen in [11]. Generally speaking, most of these methods belong to the time domain. In this section, a frequency domain called frequency domain decomposition (FDD) from [17] and [18] is discussed as an example of an ambient data algorithm. Unlike other ambient algorithms in [11] that require long time-intervals of data, FDD can provide reasonable estimates of dominant mode frequencies, damping ratios, and mode shapes in relatively short time windows, e.g., 4 min [13]. The section also discusses some important technical issues related to oscillation monitoring.

The main steps in FDD are (a) estimation of power spectrum density (PSD) matrix from PMU measurements, (b) decomposition of the PSD matrix for identification of dominant oscillatory modes in the PMU data, and (c) inverse fast Fourier transform (FFT) of the principal singular values of the PSD matrix and Prony analysis to estimate modal properties for each dominant mode. These steps are discussed in more details along with their theoretical background in the next few subsections.

PSD The high-order nonlinear power system model can be linearized around its equilibrium point and expressed in state space form as follows [3]:

$$\begin{aligned}\Delta\dot{x} &= A\Delta x + B\Delta u \\ \Delta y &= C\Delta x\end{aligned},$$

where Δx is the state vector, Δu is the input vector, and Δy is the output vector. As shown in [3], the transfer function between the input and output has the following form:

$$H(s) = \frac{\Delta y(s)}{\Delta u(s)} = \sum_{i=1}^{n} \frac{R_i}{s - \lambda_i}, \tag{3.27}$$

where $R_i = C\Phi_i \psi_i B$, Φ_i and ψ_i are right eigenvector, and left eigenvector corresponding to $\lambda_i = \alpha_i + j\omega_i$, the eigenvalue of A, respectively. And n is the order of the system model.

For a stationary random process $x(t)$, its autocorrelation function is defined as $\gamma_{xx}(\tau) = E\left[x^*(t)x(t+\tau)\right]$, where the symbol * and E stand for complex conjugate and expectation, respectively. The power density spectrum or power spectral density $S(\omega)$ is the Fourier transform of the above autocorrelation function. For a single-input single-output (SISO) system, the input and output power density spectrum has a simple relationship as $S_{yy}(\omega) = |H(\omega)|^2 S_{xx}(\omega)$. In the multi-input multi-output (MIMO) case, the relationship can be extended to the equation,

$$S_{YY}(\omega) = H(j\omega) \cdot S_{XX}(\omega) \cdot H(j\omega)^H, \tag{3.28}$$

where H denotes complex conjugate transpose.

If the inputs are represented by white noise, the power spectrum $S_{XX}(\omega)$ will be a constant diagonal matrix, denoted by F. The expression of $S_{YY}(\omega)$ for white noise inputs can be shown to be [19]

$$S_{YY}(\omega) = \sum_{i=1}^{n} \left[\frac{A_i}{j\omega - \lambda_i} - \frac{A_i^H}{j\omega - \lambda_i^*} \right] \tag{3.29}$$

where $A_i = R_i F \sum_{k=1}^{n} \frac{-R_k^H}{\lambda_i + \lambda_k^*}$.

Power Spectrum Decomposition and SVD The peaks in $S_{YY}(\omega)$ correspond to lightly damped modes of the system [17]. In the following sections, the inverse FFT of decomposed spectrum $S_{YY}(\omega)$ is shown as having an exponential form, and can be further processed by Prony analysis. This is different from post-disturbance type methods that apply Prony analysis to the PMU measurements directly. First, the theoretical derivation is originated from the following two observations when $S_{YY}(\omega)$ is evaluated near the frequency of a lightly damped mode $\lambda_r = -\alpha_r + j\omega_r$.

1. $S_{YY}(\omega)$ evaluated near ω_r can be approximated by a rank one matrix, if there is no significant contribution from other lightly damped modes nearby.
2. The rank determined by SVD of $S_{YY}(\omega)$ corresponds to the number of contributing modes when $S_{YY}(\omega)$ is evaluated near ω_r.

Mode Identification If there are no close modes nearby, $S_{YY}(\omega)$ evaluated near ω_r can be approximated by a rank one matrix [17]. Also, apply SVD to $S_{YY}(\omega)$ near ω_r, and the resulting singular values are very small except σ_1, i.e., $S_{YY}(\omega) \approx \sigma_1(\omega) u_1 u_1^H$. Thus the largest singular value σ_1 is proportional to s_r and the first left singular vector u_1 is a normalized version of mode shape w_r. Now take the inverse FFT for the largest singular value σ_1, the scaled version of s_r. Note the following Fourier transform pair $f(t) = e^{-\alpha|t|} \leftrightarrow F(\omega) = \frac{2\alpha}{\alpha^2 + \omega^2}$, and by using the frequency domain shifting property, the following is obtained:

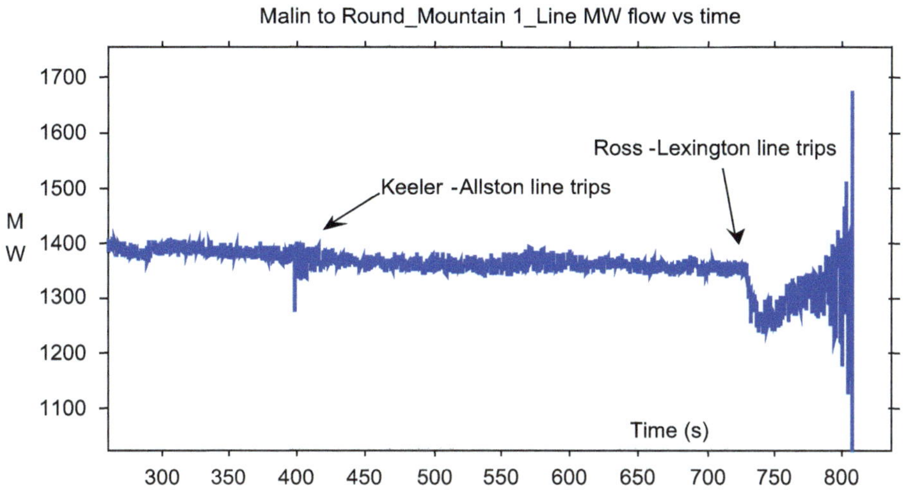

Fig. 3.21 Active power flow in the tie-line from Malin to Round Mountain line #1 during August 10, 1996 blackout [18]

$$\sigma_1(\omega) = \eta s_r(\omega) = \frac{\eta d_r}{\alpha_r^2 + (\omega - \omega_r)^2} = \frac{\eta d_r}{2\alpha_r} \cdot \frac{2\alpha_r}{\alpha_r^2 + (\omega - \omega_r)^2}$$
$$\leftrightarrow f(t) = \frac{\eta d_r}{2\alpha_r} e^{-\alpha_r |t|} e^{j\omega_r t} = B e^{-\alpha_r |t|} e^{j\omega_r t}, \tag{3.30}$$

where η is a scalar, and $B = \frac{\eta d_r}{2\alpha_r}$.

For the part $t > 0$, $f(t)$ is in an exponential form as follows:

$$f(t) = Be^{(-\alpha_r + j\omega_r)^t} = Be^{\lambda_r t}. \tag{3.31}$$

The above exponential form is processed using a simple logarithmic decrement technique for damping estimation in [17]. A recent paper [18] proposes to use the methods discussed in Sect. 3.3.2 for Prony analysis, including Prony's method, matrix pencil method, and HTLS method for damping estimation, since all these methods work well on an exponential sum model.

Case Study On August 10, 1996, a major blackout occurred in the Western American power system in the USA. At that time, a large amount of power was transmitted from the north to the south. After the loss of several transmission lines and generators [4], the system experienced growing oscillation seen across the Western American power grid. The oscillation triggered complicated relay actions in the system and finally led to a system-wide blackout. The active power in one of the major tie-lines from Malin to Round Mountain line #1 is shown in Fig. 3.21.

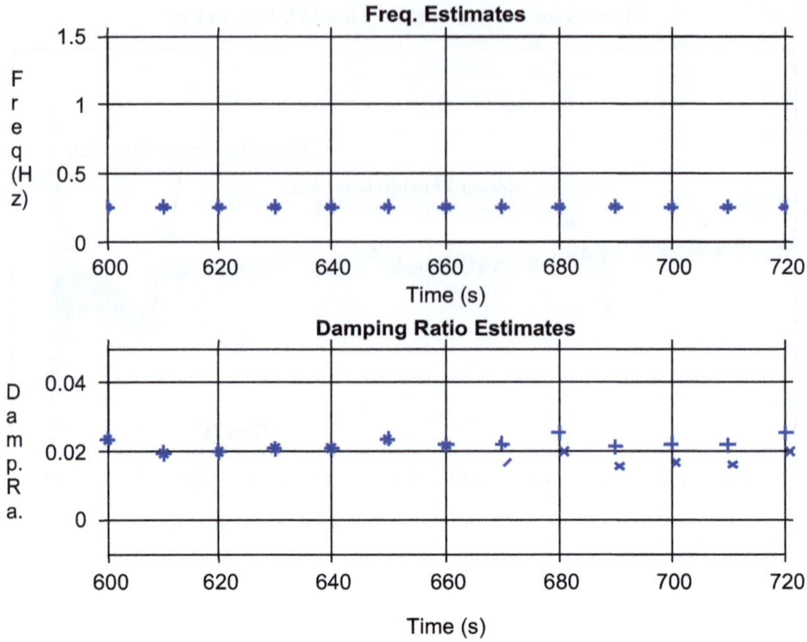

Fig. 3.22 Frequency and damping ratio estimation by FDD after Keeler-Allston line tripping [18]

After the Keeler-Allston line tripped, the system experienced a ringdown, showing poorly damped oscillations, as can be seen in Fig. 3.21. Prony analysis from Sect. 3.3.2 shows the dominant mode at 0.27 Hz with about 0.025 (or 2.5 %) damping ratio. First, FDD was applied to the data segment before the Keeler-Allston line tripping. A total of 12 signals were analyzed simultaneously. The sampling frequency was 20 Hz. Three-minute time windows of data were used for each FDD estimation. FDD analysis showed the dominant local McNary plant mode at 1 Hz with damping ratio less than 0.01 or 1 %. However, this mode did not become critical. After the Keeler-Allston line tripping, the most dominant mode changed to 0.25 Hz.

Next, the FDD was applied to the ambient data after the Keeler-Allston line tripping and before the tripping of the Ross-Lexington line (discussed as Case 0 in Sect. 3.1.2). The result in Fig. 3.22 shows the dominant mode at 0.25 Hz and the damping ratio at about 0.02 or +2 %, which matches well with the result from Prony analysis [17] of the post-disturbance data for the first tripping. The moving window analysis results show consistency, which in turn demonstrates the effectiveness of the FDD technique. The damping ratio of the 0.25 Hz inter-area mode estimated at 2 % during this time period in Fig. 3.22 could have led to operator alarms and control actions that possibly could have improved the damping levels of the 0.25 Hz in the 5-min time period between Keeler-Alston line tripping and Ross-Lexington line tripping in Fig. 3.21.

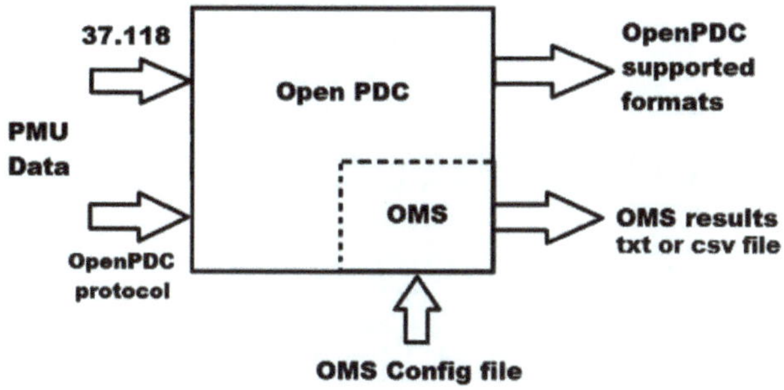

Fig. 3.23 OMS implemented within openPDC. *PMU* phasor measurement unit, *PDC* phasor data concentrator, *OMS* oscillating monitoring system

3.3.4 Real Time OMS Implementation

OMS analysis engines have been incorporated as g and Ross-Lexington line tripping in ved the damopenPDC as shown in Fig. 3.23 [12, 14]. OMS has been designed as a stand-alone oscillation analysis program that receives streaming PMU data in standard formats accepted by openPDC (including IEEE C37.118). OMS analysis results are output onto custom designed SQL server tables as part of open PDC SQL database architecture. There is an OMS configuration file that specifies user-defined OMS control parameters, such as analysis time window (say 240 s for 4 min of moving window analysis), analysis repeat rate (say 10t onto custom designed SQL server tables as part of open PDC SQL database arr tuning parameters.

OMS results include the following information for each poorly damped mode that is detected: mode frequency, mode damping ratio, mode shape information of mode shape magnitude and mode shape phase associated with each PMU, mode energy estimate, and mode estimation confidence level.

Results from the SQL server can then be processed using any visualization display program. A beta version of the visualization software developed by the project team updates the modal analysis results into a web page that authorized users can subscribe to. Figure 3.24 shows an example of a damping monitor web page captured during implementation at the utility Tennessee Valley Authority (TVA).

In the damping monitor display, there are three major areas: (1) Left top is a frequency versus damping ratio point chart. It clearly shows all the modes in the frequency domain, and their damping ratio on the y-axis. From the damping monitor result history over the previous 5 min, average values are calculated, and shown by an asterisk denoting each of the dominant modes in the power system at that time. The damping ranges over the past 5 min are displayed as white bars in Fig. 3.24. (2) The right top corner shows the time and a brief summary of the dominant modes, which includes the status of each mode. The summary will change color according to the status of each mode. For instance, the system had a poorly damped inter-area mode at 0.8 Hz with estimated damping around 2.4 % at the time when the snapshot was taken. (3) The bottom part of Fig. 3.24 summarizes mode shape information for each of the modes, which shows mode shapes of up to four modes in a radial fashion. Mode shape shows how different parts of the system are swinging against each

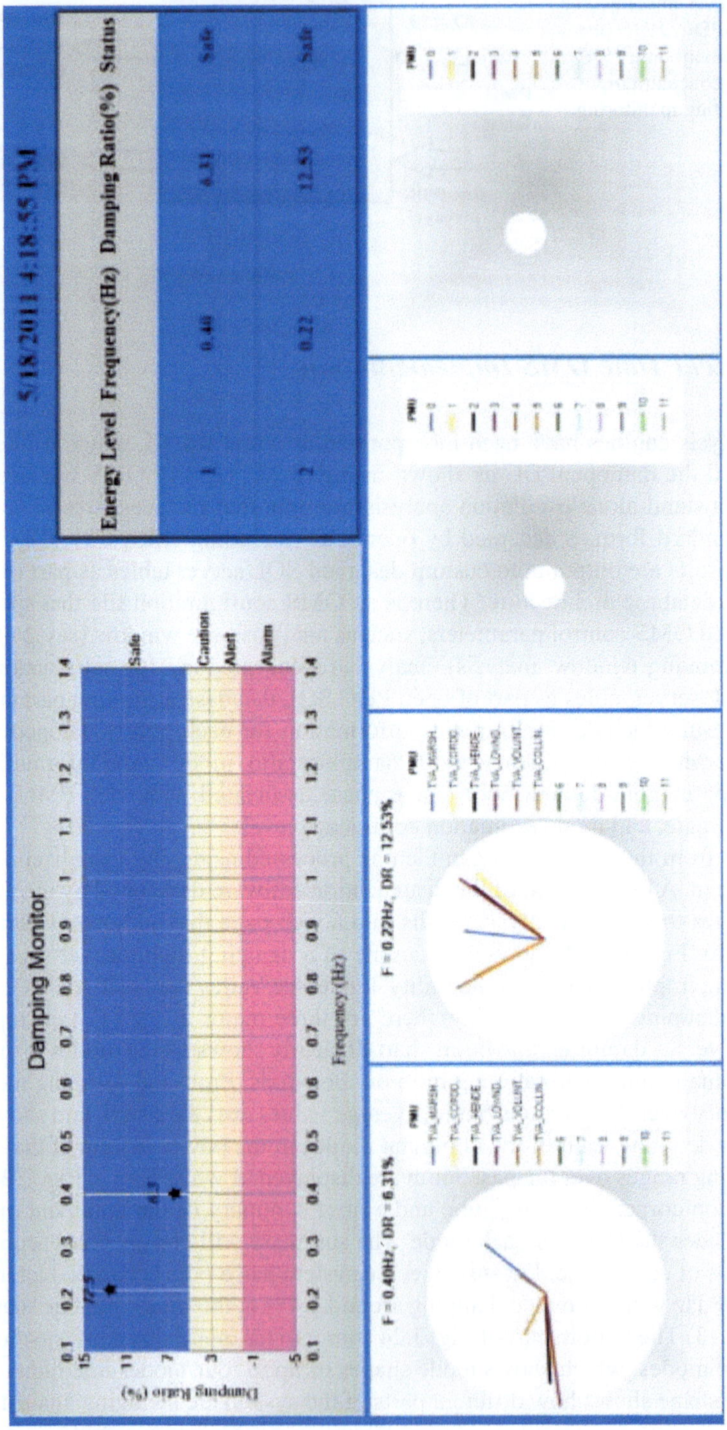

Fig. 3.24 Webpage display example at TVA [15]. *DR* damping ratio

other for each of the modes under study. This information is crucial for understanding the nature of the mode and for taking potential control actions. Different colors represent different signals, and the legends are listed on the right side.

3.4 Real Time Angle Stability Monitoring

Power systems are large interconnected nonlinear systems where system-wide instabilities or collapses can occur when the system is subjected to unusually high stress. Such system-wide blackouts lead to considerable economic costs as well as adverse societal impacts. Over the past few decades, the power-flows across distant parts of the system have been growing steadily to meet ever-increasing consumer demands. However, the construction of new transmission lines has been lagging behind because of economic as well as environmental concerns. It is well known that heavy power-flows across long transmission lines weaken the operational security of the power system with respect to oscillatory and angle instability phenomena. In essence, the steady growth of consumer demands with little new transmission support is gradually pushing the electric power system operation closer and closer towards instability limits. As a consequence, the system operation can find itself close to or outside the secure operating limits under severe contingencies.

From a technology perspective, there has been spectacular growth in the past 20 years from advances in computer and communication sciences. Together, with the emergence of the synchrophasor technology, these advances provide the opportunity for feasible and economical implementation of wide-area controls in the large electric power system. Many recent publications have analyzed the requirements and designs of wide-area controls.

One of the earliest applications of wide-area feedback control in the power system is load frequency control [3] developed in the 1970s. A secondary control loop, called the automatic generation control (AGC), coordinates individual governor responses of the generators to regulate system frequency and also to maintain the power exchanges between several control areas.

Advanced protection schemes, called special protection schemes (SPSs) or remedial action schemes (RASs), have also been developed in recent years. These schemes are designed to detect abnormal system conditions such as simultaneous loss of multiple transmission lines from appropriate relay actions, and to take predetermined corrective actions to prevent the system-wide instability. RAS schemes involve actions such as generation tripping, load shedding, capacitor insertion, or dynamic brake insertion, which are enforced at remote substations away from the fault location or other events.

Recent algorithms used in the wide-area control are based on measurements of bus voltages and generator reactive power [19]. In some applications, there is evidence to suggest that the use of phase angle measurements to detect angle stability would be more effective, especially, for monitoring the first swing instability in power systems [19]. Fast exchange of PMUs among the West Electricity

Coordination Control (WECC) utilities is also being pursued, and it is reasonable to assume the availability of system-wide phase angle information (from specific PMU locations) in the near future. This section summarizes one of the algorithms that detect and mitigate transient instability by utilizing the phase angle measurements of critical generator bus high side voltages from across the entire power system. All the simulations mentioned in this section are performed using the Transient Security Assessment Tool (TSAT) [20].

3.4.1 *Phase Angle-Based Algorithm*

An algorithm for angle instability detection has been proposed in recent papers [19, 21]. This section will discuss the algorithm along with illustrative examples on standard test systems.

Algorithm The algorithm in this section is based on the concept of the voltage-magnitude based algorithm, "Vmag," from [19] which takes into consideration the phase angle measurements. Briefly, the algorithm Vmag in [19] measures the severity of disturbances in the WECC system by quantifying the extent of bus voltage magnitude dips in the California–Oregon tie-line corridor below prespecified thresholds during system swings. The phase angle algorithm of this section analyzes the phase angles of the system from PMU measurements in a similar fashion.

The consideration of the phase angles is more challenging as compared to bus voltage magnitudes because the phase angles can vary over wide ranges during system operation. While the voltage magnitudes are kept within tight tolerances under normal system conditions, the phase angles and the relative phase differences can vary a lot across the system as determined by the tie-line MW power transfers and the availability of transmission paths for the power transfer. In order to handle this difficulty, the concept of a real time center of inertia angle reference is proposed for the computation of the system phase angle reference [21], which is then used to quantify the extent of phase angle variations away from the system center.

At present, the algorithm analyzes the phase angles in two stages: (1) the angle stability within each control area and (2) the angle stability of the entire large system. The principle in each step is similar.

First, let us recall the definition of the center of inertia angle reference (denoted COA) [3],

$$\delta_{\text{COA}} = \frac{\sum_{i=1}^{N^i} \delta_i H_i}{\sum_{i=1}^{N} H_i}, \tag{3.32}$$

where δ_i is the internal machine rotor angle for the ith machine and H_i is the respective generator inertia time constant. Since the internal machine rotor angle cannot be directly measured in the present day power system, the internal angle is approximated with the phase angle of the high side bus voltage, which is normally monitored by synchrophasors. Similarly, the inertia time constant H_i in Eq. (3.32) is difficult to access in real time because the number of units being dispatched from a generation plant can vary during operation. Therefore, the weights defined by the inertia constants in Eq. (3.32) are substituted with the high side active power injections for the generators. The machine inertias are typically proportional to the real power outputs. The modified formula, Eq. (3.33), presented below is then readily suited for real time computation using the synchrophasors.

Assume that the availability of the high side bus voltage phase angle measurements, say, δ_j^i, from a few key generating plants, say, for j from 1 to N^i in an area i. Then, the notion of the approximate center of inertia angle reference for the area, say, δ_c^i, is introduced by the rule,

$$\delta_c^i = \frac{\sum_{j=1}^{N^i} \delta_j^i P_j^i}{\sum_{j=1}^{N} P_j^i}, \tag{3.33}$$

where P_j^i denotes the current MW generation schedule at the plant j in area i. By increasing the number of phase angle measurements within each area in Eq. (3.33), the accuracy of the computation of the angle reference δ_c^i can be improved along with its redundancy. Also, this rule, Eq. (3.33), is inherently tolerant to the loss of one or more PMU channels as is commonly the case in the real time synchrophasor framework. Similarly, the center of inertia angle reference for the entire system, denoted δ_c, can be computed with the rule,

$$\delta_c = \frac{\sum_{j=1}^{N} \delta_c^i P^i}{\sum_{j=1}^{N} P^i}, \tag{3.34}$$

where N is the total number of areas that are available in the control formulation and P^i denotes the current total generation in Area i. The latest total load P^i of Area i is readily available from routine AGC calculations.

Next, a heuristic rule for detecting angle instability using these concepts is presented in a real time framework. When the representative angle δ_c^i of an area in Eq. (3.33) continuously increases away from the center of inertia δ_c beyond a pre-specified metric, this behavior would be heuristically interpreted as Area i tending to separate from the rest of the system. In this case, a suitable remedial action could be the tripping of generation in that area. Similarly, when the angle δ_c^i continues to decrease beyond a predefined threshold, it would be interpreted as a likely separation of Area i that could be countered by load shedding in Area i. These rules need to be crosschecked by analyzing the respective real time frequency measurements.

In the studies conducted, the control trigger heuristics were set to be similar to the voltage error algorithm Vmag [18]. In the case of phase angles, define $\Delta\delta_c^i = \delta_c^i - \delta_c$. Then accumulate the two integral terms, denoted as Ω_a^i and Ω_d^i, respectively, to denote the speeding up or slowing down of Area i with respect to the center of inertia reference frame. First, the term Ω_a^i is the integral for $\Delta\delta_c^i$, whenever $\Delta\delta_c^i$ continuously stays above a threshold, say $\Delta\delta_a^{i*}$. The accumulated error Ω_a^i is reset to zero whenever the angle $\Delta\delta_a^i$ drifts below $\Delta\delta_a^{i*}$. When Ω_a^i grows above a prespecified value, say Ω_a^{i*}, Area *i* is interpreted to be speeding away from the rest of the system and a suitable generation tripping may be initiated in that area. The value of Ω_a^{i*} will be tuned in real time based on the current total generation and the current spinning reserve in Area *i*. That is, the smaller the current spinning reserve (relative to the total generation) in Area *i*, the lower the threshold value for Ω_a^{i*}. The computation of the Ω_d^i is then similar to accumulating the integral of $\Delta\delta_c^i$ below a threshold, denoted $\Delta\delta_d^{i*}$. When Ω_d^i grows above a prespecified value, say Ω_d^{i*}, load shedding in Area *i* may be initiated to mitigate the potential angle instability event.

Illustration in the Two-Area System The above algorithm is implemented in the Kundur two-area system [3]. The system is simply divided into two areas with Gen 1 and Gen 2 in Area 1 and Gen 3 and Gen 4 in Area 2, respectively. Define

$$\delta_c^1 = \frac{\delta_1^1 p_{G1} + \delta_1^2 p_{G2}}{p_{G1} + p_{G2}}, \tag{3.35}$$

$$\delta_c^2 = \frac{\delta_2^3 p_{G3} + \delta_2^4 p_{G4}}{p_{G3} + p_{G4}}, \tag{3.36}$$

where $\delta_1^{\,1}$, $\delta_1^{\,2}$, $\delta_2^{\,3}$, and $\delta_2^{\,4}$ are the phase angles of the terminal bus voltages of the four generators, namely, G1 and G2 in Area 1 and G3 and G4 in Area 2, respectively. Then,

$$\delta_c = \frac{\delta_1^1 p_{G1} + \delta_1^2 p_{G2} + \delta_2^3 p_{G3} + \delta_2^4 p_{G4}}{p_{G1} + p_{G2} + p_{G3} + p_{G4}}, \tag{3.37}$$

$$\Delta\delta_c^1 = \delta_c^1 - \delta_c, \;\; \Delta\delta_c^2 = \delta_c^2 - \delta_c. \tag{3.38}$$

Apply a three-phase fault near bus 8, clear the fault and then remove three of the four lines between bus 7 and bus 8 after a certain time; the details of the simulation results are shown next. When the fault-on time is set to be 0.08, 0.10, and 0.11next. When the fau $\Delta\delta_c^1$ and $\Delta\delta_c^2$ are shown in Figs. 3.25, 3.26, and 3.27, respectively.

From the cases above, one could say that 0.10, fault-on time=0.11 s) [to be 0.08 dethree-phase fault on bus 8. Looking at Fig. 3.25, one could find that the maximum value of $\Delta\delta_c^{\,1}$ is 57.3 degrees and the minimum value of $\Delta\delta_c^{\,2}$ is −61.5 degrees for the critically stable case. Therefore, the values, $\Delta\delta_a^{\,*}=60$ degrees, and $\Delta\delta_d^{\,*}=-65$ degrees are set conservatively. Similarly, the values $\Omega_a^{\,*}=5$ and $\Omega_d^{\,*}=-5$ can be estimated from additional studies. Simulation results with different fault-on times are shown in Table 3.2. From the results, it is observed that Area 1 is moving away from the system earlier than Area 2. When some generation of Area 1 is tripped, it is found that the generation tripping action by itself is not enough to stabilize the system. Therefore, some load shedding action is initiated in Area 2 that is otherwise

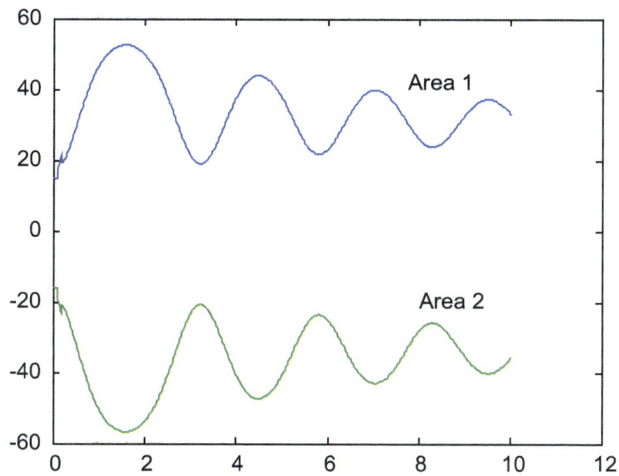

Fig. 3.25 Angles of each area (fault-on time=0.08s) [21]

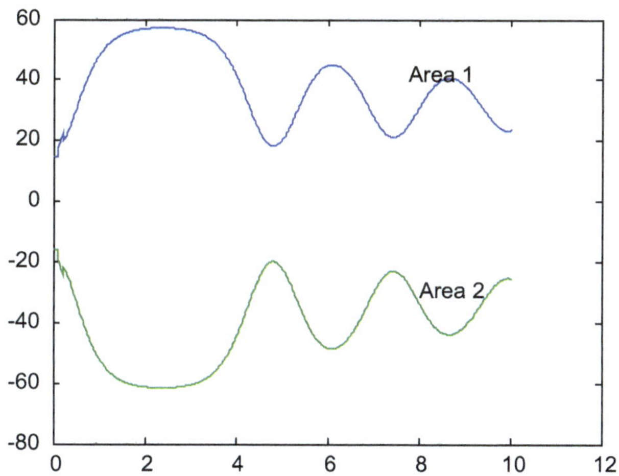

Fig. 3.26 Angles of each area (fault-on time=0.10s) [21]

decelerating from the center of the system. Gen 1 and 50ome generation of Area 1 is tripped, it is 1.83 s for the second case in Table 3.2. After the control actions, the system can be stabilized. Also, if Gen 1 is tripped at time 1.831 and 50ome generation of Area 1 is tripped, it is 1.83 s for the generation trippingia

In Table 3.2, the time duration between T_{start} (the time the phase angle goes above the threshold) and $T_{control}$ (the time the control trigger is issued) can be used to quantify the severity of the disturbance. If the time duration $T_{control} - T_{start} > T^{*}$, we can interpret the angle instability to be less severe which will be controlled by only generator tripping. Whereas if $T_{control} - T_{start} < T^{*}$, the instability can be interpreted to be more severe so that both generation tripping and load shedding will be initiated. For the two-area system, a value of say seven cycles or 0.117 s can be used for the

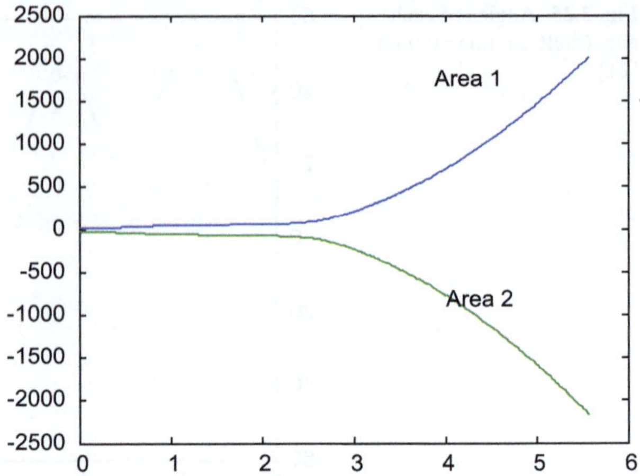

Fig. 3.27 Angles of each area (fault-on time=0.11 s) [21]

Table 3.2 Simulation results for the two-area system [21]

Fault-on time (s)	0.10	0.11		0.12	
Result	Stable	Unstable		Unstable	
Area		Area 1	Area 2	Area 1	Area 2
T_{start} (s)		1.73	1.89	1.52	1.61
$T_{control}$ (s)		1.83	1.93	1.62	1.69
Int		6.05	−5.23	6.10	−5.33
T_{unst} (s)		2.4		2.0	

Table 3.3 Potential benefits for system angle stability [21]

Fault near bus	Line removed	Critical clearing time improvement with the control (cycles)
8	7–8	2.4
7	7–8	1.8

value of T^*. In Table 3.2, T_{start} is the time$\Delta\delta_c^{\,1}$ increases beyond $\Delta\delta_c 1^*$; $T_{control}$ is the time $\Omega_a^{\,1}$ reaches $\Omega_a 1^*$; Int is the value of $\Omega_a^{\,1}$ at $T_{control}$; T_{unst} is the time $\Delta\delta_c^{\,1}$ reaches 90 degrees.

Table 3.3 summarizes the potential benefits provided by the algorithm in improving the transient stability. Taking the first case as example, the critical clearing time without the proposed control is 0.10 s (the first entry in Table 3.2). The system becomes transiently stable for the clearing time of 0.11of 0.11 the proposed control is 0.10 s (the first entry in . For thetripped, it is 1.83 s for the geneclearing time improves to 0.14 s. Compared to the 0.10 s for the original system with no control, the automatic controller as proposed provides an improved critical clearing time by a margin of 0.04 s (2.4 cycles). These simulations illustrate the fact that the controller as proposed can be useful for the two-area system in improving the transient stability of the system for critical contingencies.

3.5 Conclusions

Algorithms for real time stability monitoring of power systems using PMU measurements have been discussed in this chapter. The methodology and the examples are meant to serve as introductory material in providing a short overview of the different kinds of stability phenomena in power systems and the respective instability detection frameworks. As synchrophasors become increasingly prevalent in power systems, there will be continuing research on developing new algorithms for faster instability detection and better control designs in the near future. All the algorithms presented in this chapter assumed availability of PMU measurements in real time decision-making. In the next chapter, hybrid methods that utilize a combination of off-line model-based studies together with online PMU measurement-based decision logic will be discussed.

References

1. V. Venkatasubramanian, H. Schattler, J. Zaborszky, Dynamics of large constrained nonlinear systems: a taxonomy theory. Proc. IEEE **83**(11), 1530–1561 (1995)
2. P. Kundur, *Power System Stability and Control* (McGraw-Hill, New York, 1994)
3. M. Ilic, J. Zaborszky, *Dynamics and Control of Large Electric Power Systems* (Wiley, New York, 2000)
4. V. Venkatasubramanian, Y. Li, Analysis of 1996 Western American blackouts. *Proceeding of IREP bulk power dynamics and control-VI*, Cortina D Ampezzo, Italy, August 2004
5. K.T. Wu, D. Novosel, Voltage instability predictor (VIP)omethod and system for performing adaptive control to improve voltage stability in power systems. *US Patent Number 6219591*, 2001
6. M. Parniani, J.H. Chow, L. Vanfretti, B. Bhargava, A. Salazar, Voltage stability analysis of a multiple-infeed load center using phasor measurement data. *Proceeding of power systems conference and exposition*, 2006, pp. 1299–1305
7. V. Venkatasubramanian, X. Liu, G. Liu, Q. Zhang, M. Sherwood, Overview of wide-area stability monitoring algorithms in power systems using synchrophasors. *Proceeding of IEEE American control conference*, June 2011
8. I. Dobson, L. Liu, Voltage collapse precipitated by the immediate change in stability when generator reactive power limits are encountered. IEEE Trans. Circuits Syst. I Fundam. Theory Appl. **39**(9), 762–766 (1992)
9. X. Yue, V. Venkatasubramanian, Complementary limit induced bifurcation theorem and analysis of Q limits in power-flow studies. *Proceeding of IREP symposium on bulk power system dynamics and control-VII*, 2007
10. *Power Systems Test Case Archive*, http://www.ee.washington.edu/research/pstca/2014
11. Identification of Electromechanical Modes in Power Systems, *IEEE Task Force Report, Special Publication TP462,* IEEE Power and Energy Society, July 2012
12. J.F. Hauer, C.J. Demeure, L.L. Scharf, Initial results in Prony analysis of power system response signals. IEEE Trans. Power Syst. **5**, 80–89 (1990)
13. G. Liu, V. Venkatasubramanian, J.R. Carroll, Oscillation monitoring system using synchrophasors. *Proceeding of IEEE PES general meeting*, Calgary, 2009
14. G. Liu, J. Quintero, V. Venkatasubramanian, Oscillation monitoring system based on wide area synchrophasors in power systems. *Proceeding of IREP symposium on bulk power system dynamics and control-VII*, Charleston, 2007

15. G. Liu, Z. Tashman, V. Venkatasubramanian, P. Trachian, Oscillation monitoring system using synchrophasors. *Proceeding of IEEE PES general meeting*, 2012
16. J.W. Pierre, D.J. Trudnowski, M.K. Donnelly, Initial results in electromechanical mode identification from ambient data. IEEE Trans. Power Syst. **12**(3), 1245–1251 (1997)
17. R. Brincker, L. Zhang, P. Andersen, Modal identification from ambient responses using frequency domain decomposition. *Proceeding of the 18th international modal analysis conference (IMAC)*, San Antonio, Texas, 7–10 February 2000
18. G. Liu, V. Venkatasubramanian, Oscillation monitoring from ambient PMU measurements by frequency domain decomposition. *Proceeding of IEEE international symposium on circuits and systems*, Seattle, WA, May 2008, pp. 2821–2824
19. C.W. Taylor, D.C. Erickson, K.E. Martin, R.E. Wilson, V. Venkatasubramanian, WACS-wide-area stability and voltage control system: R&D and online demonstration. Proc. IEEE **93**(5), 892–906 (2005)
20. Powertech DSA tools, http://www.powertechlabs.com/software-modeling/dynamic-security-assessment-software/2014
21. M. Sherwood, D. Hu, V. Venkatasubramanian, Real-time detection of angle instability using synchrophasors and action principle. *Bulk power system dynamics and control-VII*, Charleston, SC, August 2007

Chapter 4
Online Transient Stability Assessment

4.1 Online Transient Stability Assessment

With the advent of deregulation in the power industry and a lack of transmission investment, today's power systems are operated much closer to their limits. Stressed system conditions mean that operators are faced with scenarios that never occurred in the past. Various critical decisions need to be taken in real time and this requires sophisticated software tools. One of the main issues that needs to be tackled deals with online dynamic security assessment and control. The objective of dynamic security assessment and control is to ensure that the system can withstand unforeseen contingencies and return to an acceptable steady state condition [1] without transient instability or voltage instability problems. In this chapter, the application of a software tool that uses artificial intelligence (decision trees; DTs) is discussed. The approach is developed by training a set of trees based on simulations that are conducted offline. With advances in computer technology, it is now possible to create and store large databases that can be used to train agents. In this case, the agents are the DTs and the DT building procedure identifies critical attributes (CAs) and parameter thresholds. In order to evaluate stability limits in terms of critical interface flows and plant generation limits, the use of synchronized measurements from phasor measurement units (PMUs) has been proposed.

The problem of online transient stability assessment has been studied extensively over the last few decades. Various methods have been suggested in order to help in maintaining system security in real time. These can be broadly classified into efficient time domain simulation algorithms, direct methods, and artificial learning approaches [2]. Figure 4.1 illustrates this classification.

M. Kezunovic et al., *Application of Time-Synchronized Measurements in Power System Transmission Networks,* Power Electronics and Power Systems,
DOI 10.1007/978-3-319-06218-1_4,

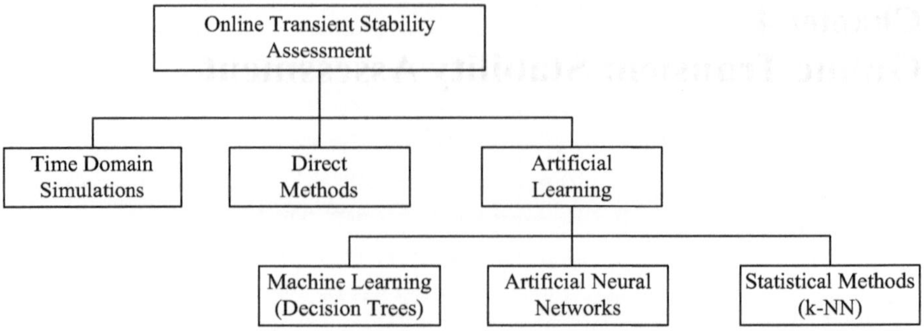

Fig. 4.1 Classification of online transient stability assessment methods

4.2 Online Transient Stability Assessment Methods

4.2.1 Traditional Time Domain Simulation

Time domain simulation is the most accurate means of determining whether a system will remain stable following a particular contingency. The differential equations of the system are solved step by step using numerical techniques in order to get the actual values of the state variables. These state variable values then yield important information regarding transient stability. Typically, the machine swing curves are computed in order to get an estimate of the maximum deviation of machine relative rotor angles. The system dynamics are simulated for the faulted and post-fault period. The post-fault period is generally of the order of a few seconds during which the system behavior is observed. However, this method is very time-consuming and does not yield any information regarding stability margins [1].

4.2.2 Direct Methods

The direct methods consist of using Lyapunov's functions in order to assess stability [2]. The transient energy function (TEF) is one such Lyapunov function that can be used to assess stability. It can be thought of as a multimachine equivalent of the more simplified equal area criterion. The basic idea consists of evaluating the kinetic energy gained by the machines during the fault period and comparing it with the maximum gain in potential energy that the system can withstand (without becoming unstable) after the fault is cleared (Fig. 4.2).

The main advantage of these direct methods is the significant reduction in computing time. Also, these methods give an idea of the stability margin of the system and hence help in evaluating the proximity of the system to instability.

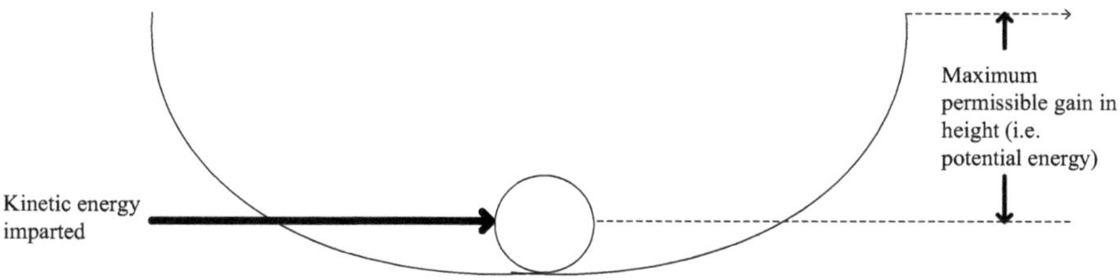

Fig. 4.2 Illustration of the TEF method

4.2.3 Artificial Learning

With advances in computer technology, coupled with the decreasing cost of mass storage devices, it has become possible to create large databases that can be used to train agents. The underlying principle in this method can be described as follows [3]:

"Given a set of preclassified examples (learning set) along with their characteristic attributes, derive a general rule (using thresholds) that can explain the input-output relationship of the preclassified cases and correctly classify new or unseen cases."

In the context of transient stability assessment, the database would consist of numerous cases that have been assessed using time domain techniques. The attributes would consist of parameters such as generator outputs, critical line flows, and relative phase angles. The output could be a stability margin or a simple classification (Secure/Insecure).

Artificial Neural Networks (ANNs) The basic building block in a neural network is a neuron or a perceptron [3]. The neuron is a linear threshold unit that can be used to model simple linear constraints. A neural network consists of many such neurons arranged in different layers. The learning process consists of identifying the structure of the network, selecting various parameters, and defining thresholds.

Statistical Pattern Recognition (K-NN) This artificial learning technique consists of matching an unseen situation with the examples stored in the database. The unseen state is assigned the class of that preclassified example whose attribute values most closely match its own. The *K* nearest neighbors (*K*-NN) method consists of classifying a state into the majority class among its *K*-NN in the stored database. Hybrid techniques that involve coupling of *K*-NN with DTs for power system transient stability assessment are discussed in [4].

Machine Learning (DTs) Machine learning is a field of automatic learning that consists of building if–then rules similar to that used by humans. This report focuses on the use of DTs for online transient stability assessment. DTs belong to the class of techniques dealing with machine learning from examples. One of the first attempts at a detailed discussion of applying DTs to assess the transient stability of a power system can be found in [5]. This chapter lays the foundation of the inductive inference method that can automatically build DTs based on a preclassified set of examples. The rules thus obtained relate the system stability to pre-fault static parameters of the system. A deductive inference method is then used to classify unseen cases in real time.

4.3 DTs for Transient Stability Assessment

Transient stability assessment deals with analysis and preventive control. Since the DT-based method uses prefault parameters in order to assess stability, it can be used as an effective approach for preventive control. The use of controllable parameters in decision rules often becomes an important aspect while taking preventive action. In this regard, stability constrained optimal rescheduling of generation is an area that has received considerable attention. Reference [6] deals with the use of the TEF method of transient stability assessment in order to carry out rescheduling of generation and critical line flows for a given initial operating condition (OC) and specified contingency. Reference [7] uses DTs as a tool for security constrained generation re-dispatch. It takes an interesting approach by using a two-step procedure. The first step consists of running an economic dispatch on all generators without considering security constraints.

$$P_T = P_1 + P_2 + P_3 + \ldots + P_i. \tag{4.1}$$

Following this, the generator set points are passed through a DT, which checks the various generator outputs. In case there are no violations, the results of the economic dispatch are displayed to the operator. However, in case one of the generators violates a decision rule, a generation re-dispatch is carried out as follows: the decision rule used in the last splitting node of the DT is used as an equality constraint in order to fix the generation of the corresponding unit. Equation (4.1) now becomes

$$P_T - P_U = P_1 + P_2 + P_3 + \ldots + P_{i-1}. \tag{4.2}$$

Thus, the total generation now available for dispatch changes to $P_T - P_U$ and this has to be distributed economically amongst $i-1$ generators. The above step is repeated iteratively till a secure solution is obtained.

The essential features of the DT transient stability method are documented and discussed in detail in [8]. Here, the selection of splitting and stop-splitting criteria used when building a DT is explained. The building of the learning set (LS) and its effects on the accuracy of the method is analyzed. Questions related to optimal size of the LS, selection of candidate attributes, and building of multiclass DTs are dealt with. Tradeoffs related to complexity versus number of classes and computational issues relating to the building and deployment of DTs in online environments are also discussed. A comprehensive discussion of the method and its critical aspects is provided. The answers provided to different basic questions about the method make it an important piece of literature in this field.

The DT transient stability method further developed and enhanced in [9] deals with the practical aspects regarding deployment. Different methods for enhancing the reliability of the DTs and for generalizing them to multicontingency stability assessment are presented.

Unlike [7–11], [12–14] tackle the problem of emergency control using DTs. Instead of using pre-fault static parameters of the system as candidate attributes for building the DT, post-fault measurements from a fault that is currently in progress are used. Using this information, the DT is used to predict if the system is moving towards transient instability. Based on the prediction appropriate emergency control action can be initiated.

4.4 Theoretical Background

4.4.1 Definition

A DT can be thought of as a flowchart representing a classification system. It consists of a sequence of simple questions regarding CAs. The answers to these questions trace a path down the tree. Figure 4.3 shows an example of a DT.

The classification made by the DT depends on the class of the terminal node. In other words, a DT can be thought of as a series of tests that are conducted on attributes of an object in order to ascertain its class.

In Fig. 4.3, A1, A2, and A3 are the CAs. For numerical attributes A1 and A2, the question compares the value with a threshold. For categorical attribute A3, the question checks to see if it belongs to a particular set. With respect to power systems, A1 and A2 can be thought of as generator outputs (numerical attributes); A3 would be the fault location or fault bus (FB) number (categorical attribute). Finally, a class is assigned to each object or case at the terminal node.

The Classification and Regression Trees (CART) software is used here to build DTs. CART is a DT procedure introduced in 1984 by Stanford statisticians Leo Breiman, Jerome Friedman, Richard Olshen, and Charles Stone [15]. A short description of the CART methodology follows.

4.4.2 CART Methodology

The procedure used by CART is called binary recursive partitioning. The term binary refers to the fact that CART splits each node into two child nodes. This splitting procedure is then applied to the child nodes, and so on. The main aspects of tree building that CART deals with are:

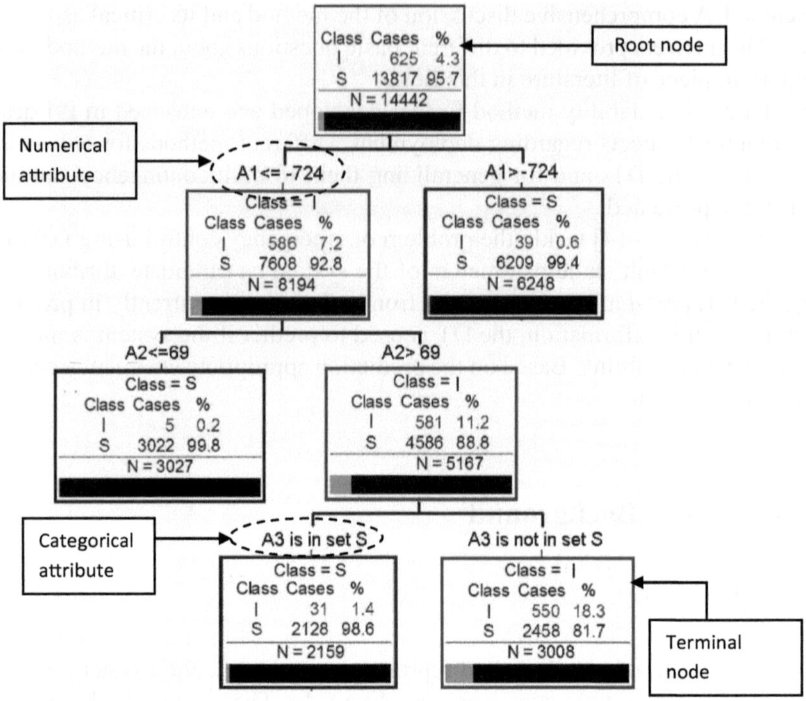

Fig. 4.3 Example of a decision tree

1. Splitting criteria
2. Deciding when to stop growing the tree

Splitting Criteria In order to split a node, CART asks questions that have a yes/no answer. Based on the answer the cases are divided into two parts. There are several questions that can be asked regarding the value of a particular attribute. Consider a simple example having 100 cases with 5 attributes. Assuming that every attribute has a unique value for each case, CART will consider 100 candidate splits for each attribute. Therefore, at each step of tree building CART evaluates 500 possible splits and chooses the best split based on certain criteria.

CART ranks candidate splits using a measure called node impurity [15]. The target variable in this case is the security of the contingency and can take on two values ("secure" or "insecure"). Each node of the tree is assigned a unique number n and will have an estimated probability ($p(S|n)$ or ($p(I|n)$) for secure and insecure cases. The Gini impurity criterion is used in this project. For node n this criterion is given by:

$$I(n) = 1 - \left[p(S|n)\right]^2 - \left[p(I|n)\right]^2. \tag{4.3}$$

If node 1 has cases with probabilities of $p(\mathrm{S}|1)=0.8$ and $p(I|1)=0.2$, then the value of the Gini diversity index is:

$$I(1)=1(0.8)^2-(0.2)^2=0.32. \tag{4.4}$$

If any node consists of only one class, then the probability of that class will be one and the probabilities for the other classes will be zero. This will result in a diversity index of zero, which suggests that the particular node has no diversity.

Using the Gini criterion, CART ranks the candidate splits and then selects the best one amongst them. The question with the highest score is selected and is called the critical splitting rule (CSR). The parameter used in the CSR is the CA. Other rules or questions that have a lower score are called competitors. Rules that mimic the action of the CSR (i.e., have the same score) are called as surrogates. A competitor can generate a good split whereas a surrogate generates exactly the same split as the CSR. As the tree grows, the nodes become more and more homogeneous.

Stop-Splitting Criteria and DT Pruning Many of the commercially available DT software packages use a goodness-of-split criterion [15] in order to decide when to stop splitting. If all the splits at a particular node do not meet the criterion, then that node is considered to be a terminal node. When all branches of the tree reach terminal nodes, the process of growing the tree is considered complete.

CART uses a different procedure in order to arrive at the best tree. It uses the data from the LS in order to build the largest possible tree. This tree is then pruned to generate several subtrees, which are tested using the data from the test set (TS). A commonly used index is the misclassification cost that is calculated using

$$R^{ts}=\frac{1}{N^{ts}}\sum_{i,j}c(i|j)\bullet N_{ij}^{ts}, \tag{4.5}$$

where N^{ts} is the total number of cases in the TS, $c(i+j)$ is the cost of misclassifying a class j case as a class i case and N_{ij}^{ts} is the number of class j cases that have been misclassified as class i. In order to avoid misclassification of insecure cases, the cost of misclassifying an insecure case as secure is higher than the cost of misclassifying a secure case as insecure (false alarm).

Unlike other DT software packages, CART does not test the splits at each node, and therefore, avoids missing important information that may be found using further splits. Finally, the DT with the least misclassification rate is selected. Other criteria, such as size or characteristics of CAs may be used to select an optimal tree.

4.5 DT Building (Preliminary Tasks)

Before building a DT for transient stability analysis the following tasks need to be carried out [8].

4.5.1 Generation of Operating Scenarios

The building of the LS is the most important aspect of the tree building process. In order to capture the essential features of the power system under study, an exhaustive list of OCs needs to be generated. Each operating point is described by its characteristic attributes, such as generation outputs, line flows, and network topology. The generation of different operating scenarios can be done by varying the generation-load profiles within a given range and/or changing network topology by simulating generator and line outages.

4.5.2 Stability Assessment

After systematically generating the operating points, various stability tests need to be performed on each operating point. A list of credible contingencies is obtained from past experience or current system status. Each contingency is then analyzed using transient stability assessment software and the results (i.e., "secure" or "insecure") are stored. This contingency analysis is carried out for all the generated operating points.

4.5.3 Selection of Attributes

The selection of candidate attributes should be done such that they adequately characterize the properties of the system. Based on past experience, certain parameters such as generator power (real and reactive) outputs, power flows on critical lines, phase angle differences, and information about network topology can be used. Since the DT would be used for control purposes, the attributes selected should be controllable. Selection of simple controllable attributes will yield trees that are easy to interpret, whereas the use of more complex attributes will result in trees that are more accurate and suitable for the purpose of overall system analysis.

4.5.4 Preparation of Database

The selected attributes and results of the stability analysis for each contingency (for a particular OC) are systematically stored in the database. This step is then repeated for all the OCs. Therefore, each case is characterized by a certain number of CAs and a predetermined (from stability analysis) stability class. Table 4.1 shows a sample database. This set of cases is then separated into a LS and TS. The LS is used to train the DT, and the TS is used to evaluate the performance of the DT on unseen cases. Generally, a randomly selected fraction of the entire database is used as the TS (20–30% of all cases).

Table 4.1 Sample database

Class	Fault bus	Other bus	P_98537_98606 (MW)	PG_98233 (MW)	A_98606 (deg)
Secure	97745	97746	45	100	15
Insecure	97747	97750	60.5	100	17
Secure	98654	98656	63	110	18
Secure	98634	98632	50	110	17
Insecure	98767	98...	...	...	...
...	...	...	...		

4.5.5 Multiclass and Multicontingency DT Considerations

The number of stability classes to be considered while building the DT needs to be decided in advance. It should be kept in mind that the LS should have an adequate representation of all the considered stability classes. The DT performance depends on the quality of the LS, and if sufficient numbers of cases belonging to every stability class are not present in the database, the accuracy and reliability of the DT in classifying unseen cases will be poor. Also, DTs can be built for a single contingency or for multiple contingencies. In case of multicontingency DTs, attributes related to the fault location and type of fault can be used as predictors. The multicontingency DTs will give a list of contingencies that will result in instability for a given OC. On the other hand, a single contingency DT will provide information about which OC will make the system unstable in the event that the particular contingency (for which the DT is built) occurs.

4.6 Use of PMUs for Transient Stability Assessment

The system operator is generally provided with a set of operating guidelines that need to be followed in order to maintain secure system operation. These guidelines are obtained by carrying out extensive offline studies. Critical parameters are identified, and thresholds for these parameters are also evaluated. The operator is then required to maintain system operation in a secure region based on these constraints. Traditionally, operators rely on SCADA measurements and the state estimator in order to ensure that the system is operating within the secure region. However, since the values obtained from SCADA measurements and state estimators are not synchronized, it is difficult to accurately measure the current OC.

The main advantage of using PMUs is that they provide GPS-synchronized measurements that can then be used to calculate phase angles, frequency, and rate of change of frequency. The time synchronism that can be achieved with the use of PMUs makes it possible to obtain a snapshot of the system condition at any given point in time.

Operational nomograms are built offline in order to assess stability in real time. If the current OC falls in a particular region, then an accurate assessment of the system stability can be made using the stored nomograms. The effectiveness of this method depends upon the accuracy and reliability with which OCs can be measured. This is where PMUs can be of great help. With accurate synchronized measurements of phase angles and critical power flows throughout the system, PMUs can give system operators the precise information needed in order to characterize the current operating point.

4.7 DT-Based Online Security Assessment Scheme

A flow chart describing the DT-based online security assessment scheme is given in Fig. 4.4. There are three main steps involved in building the DT and using it for online security assessment.

The offline DT building consists of creating a large database of cases along with their classification using time domain simulations. This database is then used to build a DT. Depending on the system conditions, the DT built offline can be updated on an hourly basis. The updated DT is then used online for transient stability prediction with the help of accurate synchronized measurements from PMUs.

4.7.1 Offline DT Building

This stage is executed offline to generate a database of cases and build a DT for a 24-h (day ahead) horizon. Initially, N_{OC} prospective OCs for the next 24 h are obtained from short-term load forecast and unit commitment programs. These OCs are selected such that they adequately represent probable system behavior with regard to generation and load profiles and network topology. N_C contingencies based on type, location, and fault duration are either assumed throughout the power system or selected by the operator from a history of critical contingencies. For each OC, detailed time-domain simulations of all the N_C contingencies are conducted. Specified security criteria dealing with transient stability, transient voltage dip, transient frequencies, and damping ratios are then checked to determine the security classification for each case. The classifications may be binary (i.e., secure or insecure) or multiple, e.g., transient instability, insufficient damping, voltage insecurity, and frequency insecurity, with different priorities assigned to each criterion. Thus, the class of an insecure case is determined by the criterion with the highest priority. Finally, a database of $N_C{}'N_{OC}$ cases, each including a security classification and a vector of attribute values, is generated and further divided into an LS and a TS.

A group of predictors is then selected from either fault-dependent (FD) attributes, e.g., the fault type and location, or fault-independent (FI) attributes, e.g., bus voltage angles, MW transfers across lines or interfaces, outputs of generators.

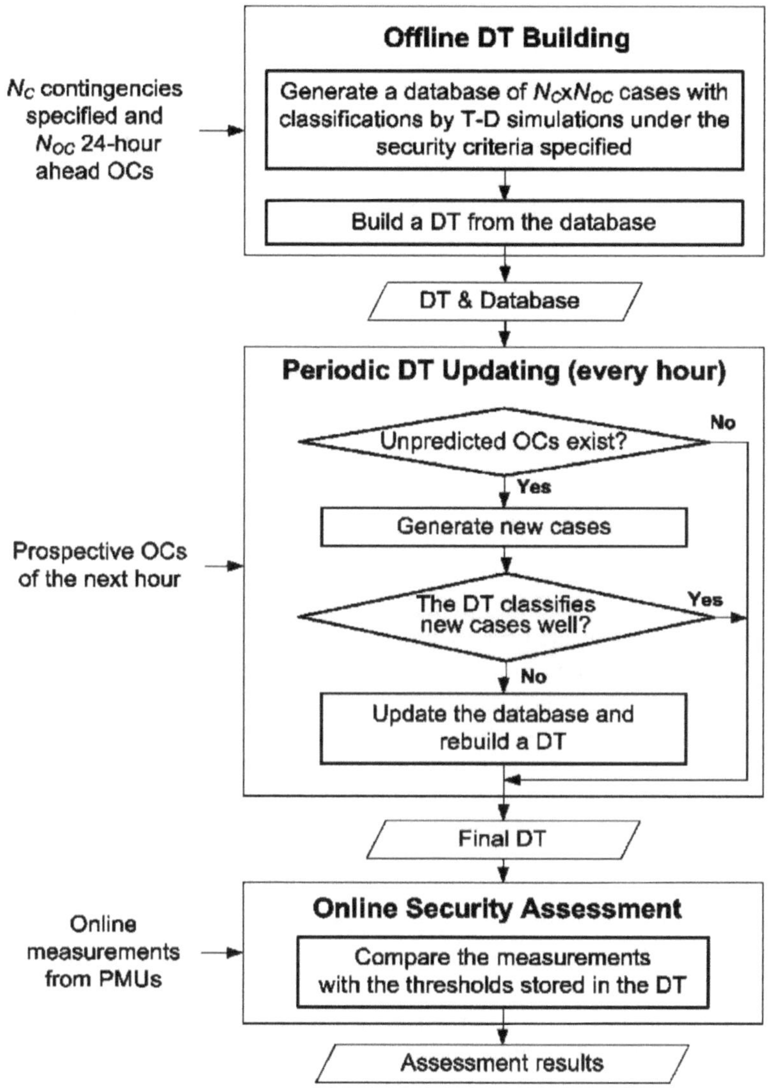

Fig. 4.4 Flowchart describing the proposed scheme

Some predictors critical to the system's security are screened out as CAs to form the CSRs of the DT. Fault-independent critical splitting rules (FI-CSRs) are of particular interest because they define the insecure regions in the attribute space where the current OC can be located. To determine the exact location of the OC, it is generally necessary to obtain accurate synchronized values of fault-independent critical attributes (FI-CAs). A number of PMUs have been installed in many power systems and they provide accurately synchronized measurements of frequently changing parameters, including the voltages of the buses where they are installed and the currents and power flows of the branches connected to the bus. If impedances of the branches are constant, then voltage angles at the other ends of the branches can also be acquired. Thus, the synchronized measurements from PMUs are good candidates for FI predictors. In addition, predictors can be selected from the parameters monitored by the SCADA system if they do not change frequently, e.g., outputs of generators. In this case, the measurements obtained need not be synchronized.

Following this, classification tree algorithms are used to build a DT from the learning and TSs. CAs and CSRs are selected from all available attributes and are stored in each inner node. The DT built could be two-class (insecure or secure) or contain $N + \mathrm{e}$ classes if N insecure classes are considered. If storage space allows, even N two-class DTs could be built at the same time, each of which focuses on a different type of insecurity.

4.7.2 Periodic OC Prediction and DT Updating

The time horizon is divided into periods of equal length (typically 1 h). The DT update procedure is executed for each period in order to predict the performance of the DT for the upcoming time period. In case the performance of the DT is not satisfactory, a new DT can be built.

Prospective OCs in the next period can be predicted using a short-term load forecast (this could be more accurate than the forecast used previously) and unit commitment rules. If the prospective OCs are close to any of the N_{OC} OCs already considered, the DT will remain unchanged until the next period. If new OCs appear, then the DT may not perform well. Thus, the N_{C} contingencies will be simulated again at the new OCs to generate new cases, which are used to test the existing DT. If the DT performance is still satisfactory, the DT is retained until the next period in the time horizon. Otherwise, the new cases together with the old are used to build a new DT.

The speed of building a DT from scratch is discussed below. For the case studied in Sect. 4.11, 280 "$N-1$" contingencies are considered for 56 OCs of the online EMS representation of the Entergy system. Thus, 15,680 simulations are needed. Using a PC with a single Pentium-4 3.4 GHz CPU, each simulation takes 5–10 s, and the 280 simulations for each OC are completed within 50 min. Since all simulations are independent, they can be executed on a system with multiple parallel CPUs to accelerate the computation time. Finally, CART's growing and pruning processes are extremely fast such that an iterative DT building procedure may be designed to quickly generate a good DT from the simulation results.

When compared to building a new DT, updating an existing DT takes much less time. This is because the simulations need to be carried out for new OCs only, and a DT needs to be built from the updated database.

4.7.3 Online Security Assessment

In real time, the control center obtains synchronized measurements of various parameters throughout the system. If the FI-CAs used to build the DT are selected from existing PMU measurements, then the operator can perform accurate dynamic security assessment for a single contingency or a group of contingencies. To consider a general case, assume that N'_C ($1 \leq N'_C \leq N_C$) contingencies are considered (N_C denotes the total number of contingencies considered during simulation). The following procedure can predict security and give guidelines for preventive control:

- Prune all branches of the DT that do not correspond to the contingencies of interest. If the operator is only interested in faults at a single bus (say "Bus 1"), then only that branch of the DT dealing with "Bus 1" needs to be considered. In such cases, the use of FD CAs in the initial splits can prove to be helpful.
- Use operational nomograms (constructed offline) to ascertain the location of the current OC in the attribute space.
- If the DT predicts an insecure condition following a particular contingency or set of contingencies, arm appropriate preventive control measures.

4.8 Reliability of the DT

When using the DT constructed for online security assessment, it is necessary to understand the importance of each predictor. The reliability of a particular CSR depends on the number of cases in the node where it is used. A CSR for a parent node with a large number of cases is obviously more reliable than a CSR for a parent node with a smaller number of cases. Therefore, the first predictor carries the highest weight since it is derived by observing all the cases in the LS. The predictors used in the later stages of the DT operate on a certain fraction of the data only and hence carry less weight.

When OCs change, it is likely that the predictors used near the terminal nodes will lose their credibility. In such cases, it becomes necessary to use a path-based approach in order to ascertain the class of a terminal node. In the normal DT building procedure, the class of the terminal node is assigned based on the proportion of secure/insecure cases in the terminal node. However, this does not take into account the path followed by a particular case before entering the terminal node. In order to account for the path traced by any case as it drops down the tree, an insecurity score (IS) can be calculated for each path. The IS calculates the weighted average of the percentage of insecure cases in the nodes belonging to each path.

$$\mathrm{IS} = \frac{\sum_{i=1}^{K} p_i \times \omega_i}{\sum_{i=1}^{K} \omega_i}, \tag{4.6}$$

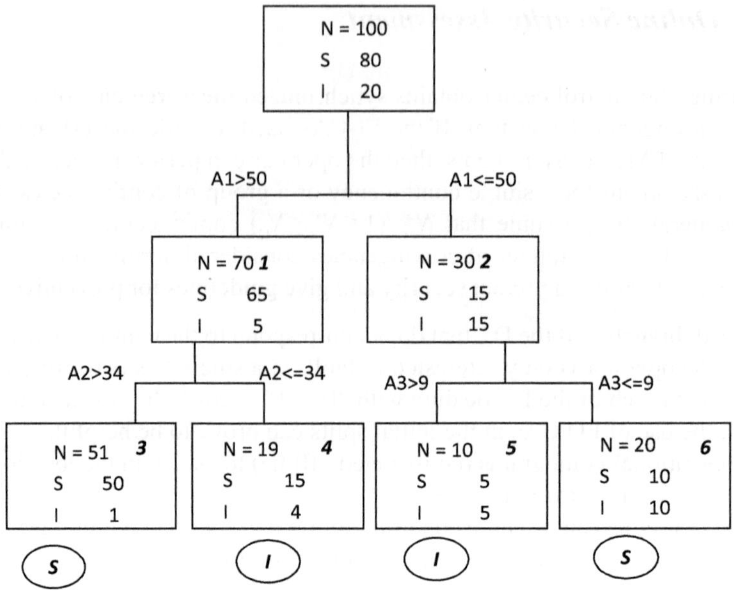

Fig. 4.5 Sample decision tree (*DT*) illustrating new path-based approach. *S* secure, *I* insecure

Table 4.2 ISs of paths for sample tree

Path	Insecurity score	Predicted class (path based)	Predicted class (terminal node based)	Secure/insecure
1–3	4.96	S	S	50/1
1–4	10.11	S	I	15/4
2–5	50.00	I	I	5/5
2–6	50.00	I	S	10/10

S secure, *I* insecure

where

p_i percentage of insecure cases in node *i* and
ω_i weight assigned to node *i*.

In Fig. 4.5, the IS for path 1–3 equals the weighted average of the percentage of insecure cases in nodes 1 and 3. In order to classify a terminal node the IS of the path leading to that node is calculated and compared to a threshold.

A sample set of ISs are calculated for the tree in Fig. 4.5. The threshold for the IS used here is 30 %. All paths with scores greater than 30 % are considered to be insecure (Table 4.2).

The prediction accuracy using the traditional terminal node-based method (Table 4.3) and the new path-based method (Table 4.4) is compared below.

Table 4.3 Prediction accuracy for traditional method

Actual class	Total cases	Percent correct	S $N=71$	I $N=29$
S	80	75	60	20
I	20	45	11	9

S secure, I insecure

Table 4.4 Prediction accuracy for path-based method

Actual class	Total cases	Percent correct	S $N=70$	I $N=30$
S	80	81.25	65	15
I	20	75	5	15

S secure, I insecure

Using the path-based method, better results are obtained for both secure and insecure cases. The path-based classification approach increases the robustness of the DT to variation in OCs. This method when used on the Entergy system yields promising results.

4.9 Preventive Control

After determining the contingencies that will result in instability (for a particular OC), appropriate preventive control action must be initiated if these contingencies occur. Using the DT, the CAs whose thresholds are being violated can be determined. In Fig. 4.5, $A1=50$ is a threshold value. In order to make more cases enter node 1, the value of A1 can be made greater than 50. However, this approach does not always yield good results.

Although the thresholds for the critical parameters are accurate to classify system security in real time, they need not be effective for preventive control. The DT in Fig. 4.5 suggests that $A1 \leq 50$ results in insecurity. Assume that attribute A1 denotes the power flow through a line. Increasing the power flow of the line beyond 50 will not necessarily result in a secure OC. Given the fact that the power flow through a line can be changed in many different ways, the resulting OC might be a new one that has not been considered in the database. Therefore, when designing preventive control measures, it is necessary to ensure transition to a secure OC that has already been analyzed and proven to be secure.

4.10 Building a Better DT by Adjusting Penalties for Predictors

Both FI-CAs and FD-CAs are used in the DT. Sometimes, FI-CAs are preferable since they are measurable and partially controllable. For the purpose of security analysis, it may be important to separate or change the order in which they are used in the splitting process. FD-CSRs could be used before FI-CSRs to identify insecure areas of the system (by location), or alternatively, FI-CSRs could be used before FD-CSRs to identify insecure operating regions in the attribute space and design preventive control. Either approach can be used, by assigning appropriate penalties between 0 and 1 to FI or FD predictors in order to reduce their improvement scores. A penalty of 0 and 1, respectively, means "no limitation on using" and "not using" the predictor. The higher the penalty a predictor has, the later it is used in the DT. The following procedure can be used to build a DT that uses FD-CSRs after FI-CSRs by adjusting the penalties of FD predictors.

Procedure 1:

1. Assign each FD predictor a penalty $p = p_{min}$.
2. Generate a series of DTs by CART's growing and pruning processes and select the best DT. The best DT is the one with the lowest misclassification cost. The child nodes formed as a result of using FD-CSRs are considered as the terminal nodes and the rest of the tree is pruned. Thus, a DT using FD-CSRs only at the last split is generated and stored.
3. Let $p = p + \Delta p$, if $p \geq p_{max}$, select the best DT from all DTs stored and end the procedure; otherwise, go to step 2.

Heuristic techniques can be applied to find a good value for the penalty. Similar procedures can be designed to generate a DT that meets other requirements.

Another approach to build DTs is to only select predictors from critical variables e.g., plant generation outputs of critical plants, power flows of critical interfaces or transmission lines, and phase angle differences between critical buses. It is important to note that given a set of predictors, CART's DT building algorithm does not guarantee building an optimal DT. At each step of DT growing, i.e., splitting a node, the algorithm simply seeks the CSR that best separates cases of different classes only for that particular node. CART does not take into account the effect that the current split will have on the overall classification accuracy. CART generates a locally optimal split at each stage as opposed to a globally optimal split. Simulations have shown that replacing a CSR with a competitor could totally change the nature of the DT and result in better prediction for unseen cases. Finding an optimal DT given a set of predictors is a complicated problem, and results in a very large number of predictor combinations. However, the following Procedure 2 can be used to quickly find a good DT by recursively executing CART's DT building algorithm.

Procedure 2:

1. Select a group of reasonable predictors and build a DT. Let $i = 1$.
2. For the CSRs of all nodes at the ith level starting from the root node, check their nonsurrogate competitors.
3. If a CSR at a node has a good nonsurrogate competitor whose improvement score is larger than γ times (e.g., $\gamma = 0.9$) of its score, then assign its associated predictor the minimum penalty that ensures that the new DT uses the competitor instead of the old CSR at that node. The value of the "minimum penalty" is easily determined by a simple trial and error procedure. Then, go to step 5.

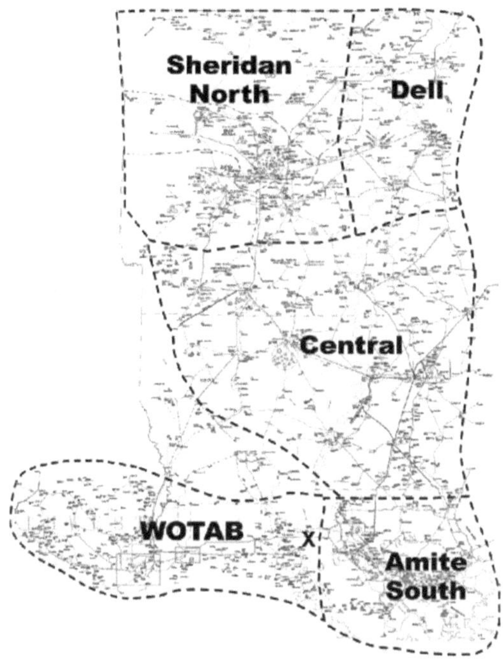

Fig. 4.6 Entergy transmission system map

4. If no such good nonsurrogate competitor exists at the *i*th level, go to step 6.
5. If the new DT built in step 3 is better than the old DT, replace the old DT by the new DT and go to step 2.
6. If no parent node exists below the *i*th level, then end the procedure; otherwise, let $i = i + 1$ and go to step 2.

Using the above procedure, an optimal DT for a given set of predictors can be generated. A better understanding of the concepts described in this section can be gained through the case studies on the actual Entergy system in Sect. 4.11.

4.11 Case Study on Entergy System

The proposed scheme was tested on the Entergy online operational model, which consists of around 240 generators, 2000 buses, and 2100 lines. In order to carry out the analysis in a systematic manner, the entire network was divided into five areas as shown in Fig. 4.6. These areas were chosen based on the information provided by Entergy regarding its system configuration. This case study focuses on the Amite South area, with 54 generators, about 400 buses, and approximately 430 lines, and includes the city of New Orleans. A DT is built for a 24-h horizon. The different aspects of DT building and analysis are dealt with in detail in the next few sections of this chapter.

4.12 Database Generation

4.12.1 Operating Conditions

The power flow data for 2 days was obtained from Entergy. The data consist of light load and heavy load power flow profiles for July 19, 2006 and July 26, 2006. The network topology and generation-load profile are considered on a day-ahead basis for 24 h of July 19, 2006. A credible stressed situation with a 500 kV line from Wells to Webre being out of service is considered. This is shown by the cross ("X") in Figs. 4.1 and 4.2. The Wells-Webre line connects the WOTAB area and Amite South area. Using this as the starting point, the light load (17,819 MW) and heavy load (26,600 MW) power flow profiles of July 19, 2006 were used to generate 54 intermediate OCs. Percentage scaling was used to vary the Entergy load, and the generation was adjusted using certain economic dispatch rules in order to achieve generation load balance. Generators were switched in and out of service depending on a list provided by Entergy. While changing generation, it was ensured that minimum and maximum generation limits were not violated.

4.12.2 Contingencies

For each of the 56 OCs the following contingencies in the Amite South area were considered:

- Three phase faults on both ends of all 500 kV lines with a fault clearing time of five cycles
- Three phase faults on both ends of all 230 kV lines with a fault clearing time of six cycles

There are a total of 280 such contingencies for each OC. This results in a total of (280 × 56) 15,680 cases. Twenty percent (20 %) of these cases (0.20 × 15,680 = 3136) are used as the TS and the rest (12,544) of the cases are included in the LS. Each case is simulated in Powertech Lab's TSAT software [16], and the two security criteria, namely, transient stability and damping are checked. Different DTs are built for transient stability and damping problems using the CART software developed by Salford Systems [17]. The criteria used are as follows:

1. Transient stability criterion: After the fault is cleared, the transient stability margin of the system should be greater than 5 % and no bus voltage should violate the voltage range of 0.70–1.20 p.u. for more than 20 cycles. Here, the stability margin is estimated by TSAT's power swing-based algorithm [16].
2. Low damping criterion: Entergy requires the damping ratio to be greater than 3 %. This criterion is applied to the inter-area oscillation modes of generator rotor angles (frequency range is 0.25–1.0 Hz).

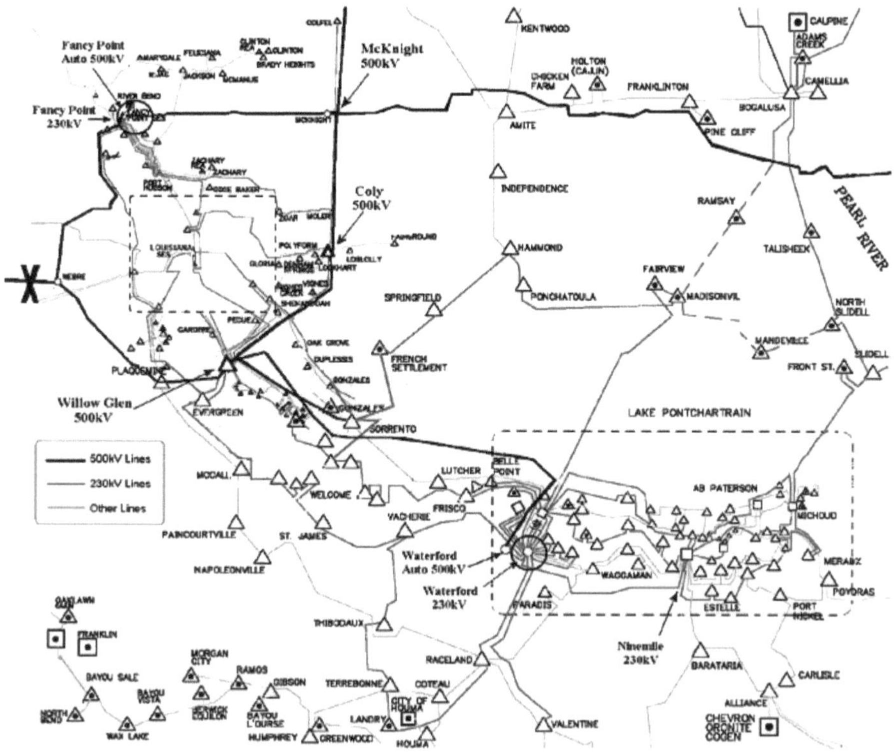

Fig. 4.7 Amite South area (showing phasor measurement unit (*PMU*) locations)

4.12.3 *Candidate Attributes*

Entergy has placed 9 PMUs to monitor critical interfaces in order to get good visibility of the system. The Amite South area has 3 PMUs at buses Waterford 230 kV and Fancy Point 230 kV (encircled in Fig. 4.7). The PMUs measure voltages of the buses and currents of the following lines:

1. Waterford 230 kV (bus 98537) looking at line to Ninemile 230 kV (bus 98606)
2. Waterford 230 kV (bus 98537) looking at Waterford Auto 500 kV (bus 98539)
3. Fancy Point 230 kV (bus 98234) looking at Fancy Point Auto 500 kV (bus 98233)

Therefore, the voltage phase angles at bus numbers 98233, 98234, 98537, 98539, and 98606 can be obtained from the PMU measurements. Also the line flows of the three lines mentioned above are available from the PMUs. In order to obtain additional information regarding the potential locations of new PMUs, certain additional measurements from other 500 kV buses in the area are also included in the list of CAs. There are a total of eight other 500 kV buses in the area. Two groups of predictors are used to build the DTs.

Group 1 (from candidate PMU measurements):

1. The serial number of the FB
2. MW-flows of the 16 branches (each denoted by *P_i_j*, where *i* and *j* are two bus serial numbers)
3. Angle differences between voltage phase angles of 13 buses (each denoted by *A_i_j*, where *i* and *j* are two bus serial numbers)

Group 2 (from existing PMU measurements):

1. The serial number *FB*
2. MW-flows of the three branches where PMUs are currently installed (P_98537_98606, P_98537_98539, and P_98233_98234)
3. Angle differences between voltage phase angles of five buses currently being monitored by PMUs (98233, 98234, 98537, 98539, 98606)

4.13 Results for Entergy System

The classification accuracy obtained from each group of predictors for DTs built for the transient stability and damping criteria are as follows:

4.13.1 DTs for Transient Stability (DT-U)

Group 1 (candidate PMU measurements, DT-U1) (Fig. 4.8 and Tables 4.5–4.6)

Group 2 (actual PMU measurements, DT-U2) (Fig. 4.9 and Tables 4.7–4.8)

4.13.2 DTs for Damping (DT-D)

Group 1 (candidate PMU measurements, DT-D1) (Table 4.9 and 4.10)

Group 2 (actual PMU measurements, DT-D2) (Table 4.11 and 4.12)

The predictors denoting line flows used by DT-D1 and DT-D2 are given by P_98246_98390, P_98246_98539=P_98539_98537, and P_98537_98606 (Fig. 4.11). Note that although Fig. 4.10 shows P_97301_98430 as a predictor, the power flows of the key transmission path shown in Fig. 4.12 generate equally good splits. Using the thresholds for the CAs identified in DT-D2, the following nomogram is built (Fig. 4.13).

The numbers written in the various regions denote the probability that an OC lying in the region is insecure. The darker shades depict regions of high insecurity. This nomogram can be used in real time to evaluate the state of the system and arm appropriate control action. For example, if the operator finds that

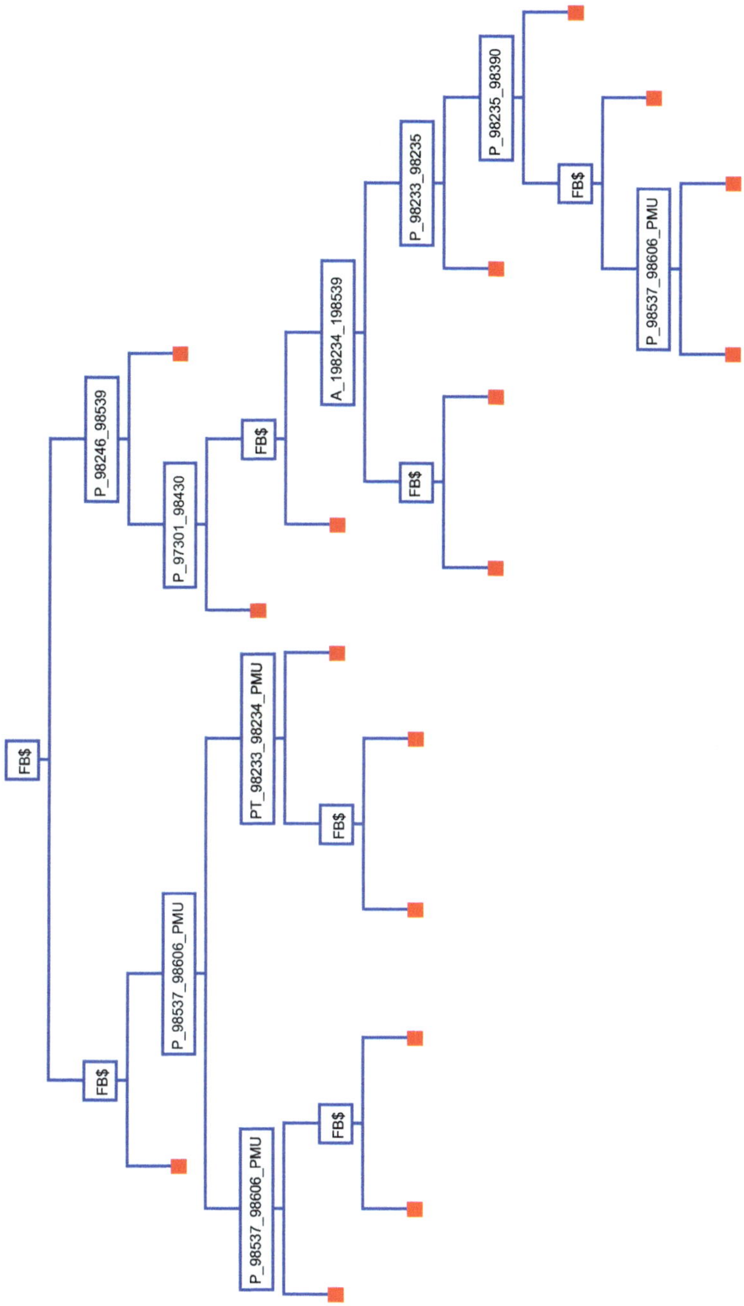

Fig. 4.8 Details of predictors used at each stage of the DT-U1

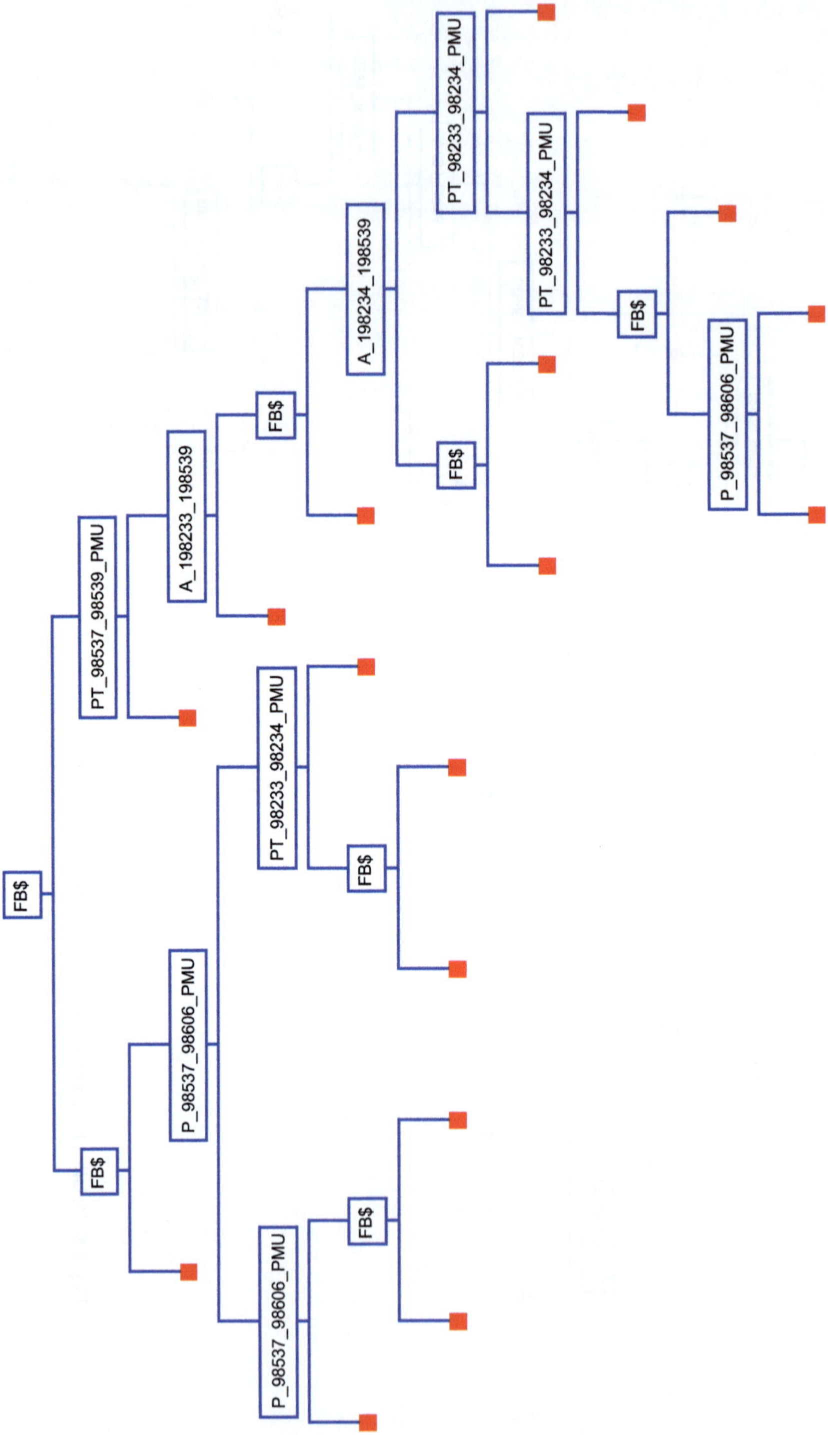

Fig. 4.9 Details of predictors used at each stage of DT-U2

Table 4.5 Prediction accuracy for learning set (DT-U1)

Actual class	Total cases	Percent correct	Percent misclassified	S N=11,722	2
S	12,263	95.59	4.41	11,722	541
U	281	100.00	0.00	0	281

S Stable, *U* Unstable

Table 4.6 Prediction accuracy for test set (DT-U1)

Actual class	Total cases	Percent correct	Percent misclassified	S N=2943	U N=193
S	3062	96.05	3.95	2941	121
U	74	97.30	2.70	2	72

S Stable, *U* Unstable

Table 4.7 Prediction accuracy for learning set (DT-U2)

Actual class	Total cases	Percent correct	Percent misclassified	S N=11,722	U N=822
S	12,263	95.59	4.41	11,722	541
U	281	100.00	0.00	0	281

S Stable, *U* Unstable

Table 4.8 Prediction accuracy for test set (DT-U2)

Actual class	Total cases	Percent correct	Percent misclassified	S N=2943	U N=193
S	3062	96.05	3.95	2941	121
U	74	97.30	2.70	2	72

S Stable, *U* Unstable

Table 4.9 Prediction accuracy for learning set (DT-D1)

Actual class	Total cases	Percent correct	Percent misclassified	D N=2991	S N=9553
D	2015	99.45	0.55	2004	11
S	10,529	90.63	9.37	987	9542

D Damping, *S* Stable

Table 4.10 Prediction accuracy for test set (DT-D1)

Actual class	Total cases	Percent correct	Percent misclassified	D N=715	S N=2421
D	486	97.94	2.06	476	10
S	2650	90.98	9.02	239	2411

D Damping, *S* Stable

Table 4.11 Prediction accuracy for learning set (DT-D2)

Actual class	Total cases	Percent correct	Percent misclassified	D N=2921	S N=9623
D	2015	98.96	1.04	1994	21
S	10,529	91.20	8.80	927	9602

D Damping, *S* Stable

Table 4.12 Prediction accuracy for test set (DT-D2)

Actual class	Total cases	Percent correct	Percent misclassified	D N=705	S N=2431
D	486	96.71	3.29	470	16
S	2650	91.13	8.87	235	2415

D Damping, *S* Stable

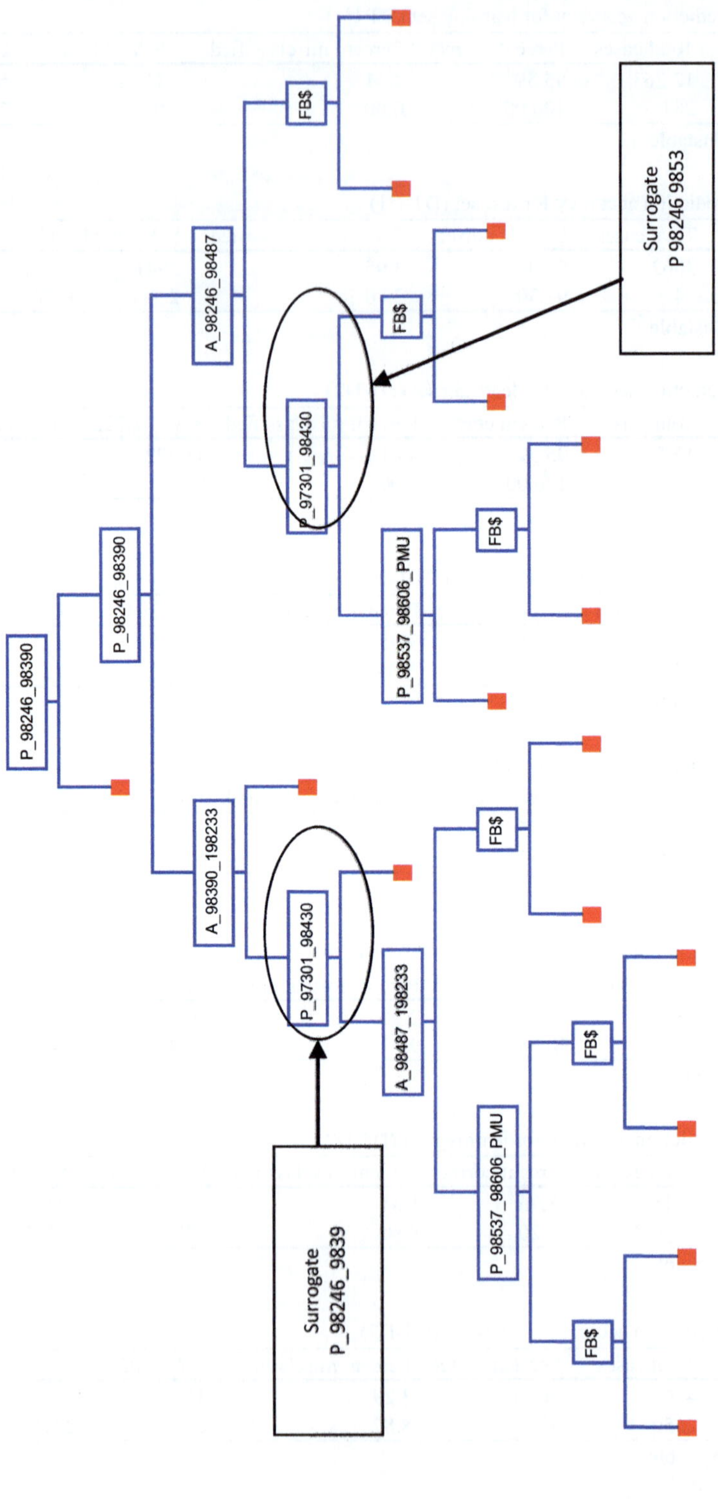

Fig. 4.10 Details of predictors used at each stage of DT-D1

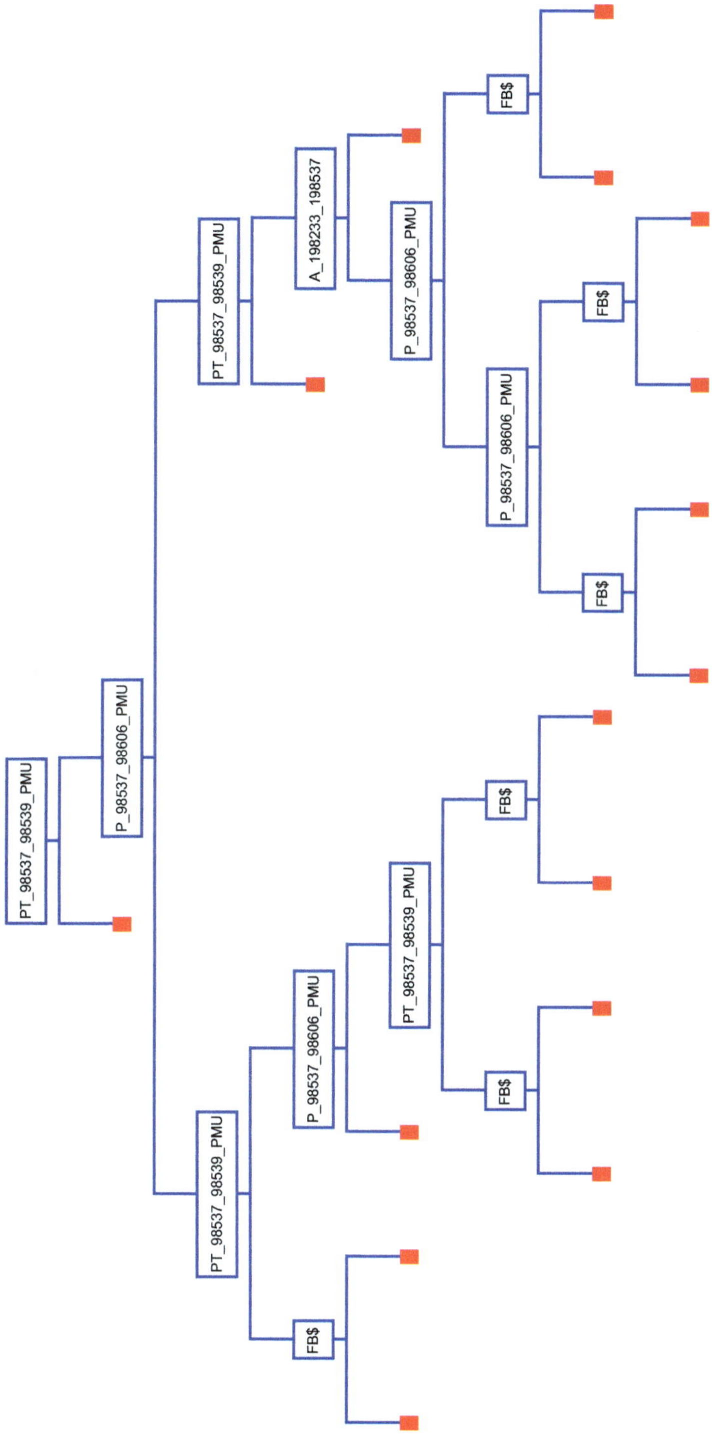

Fig. 4.11 Details of predictors used at each stage of DT-D2

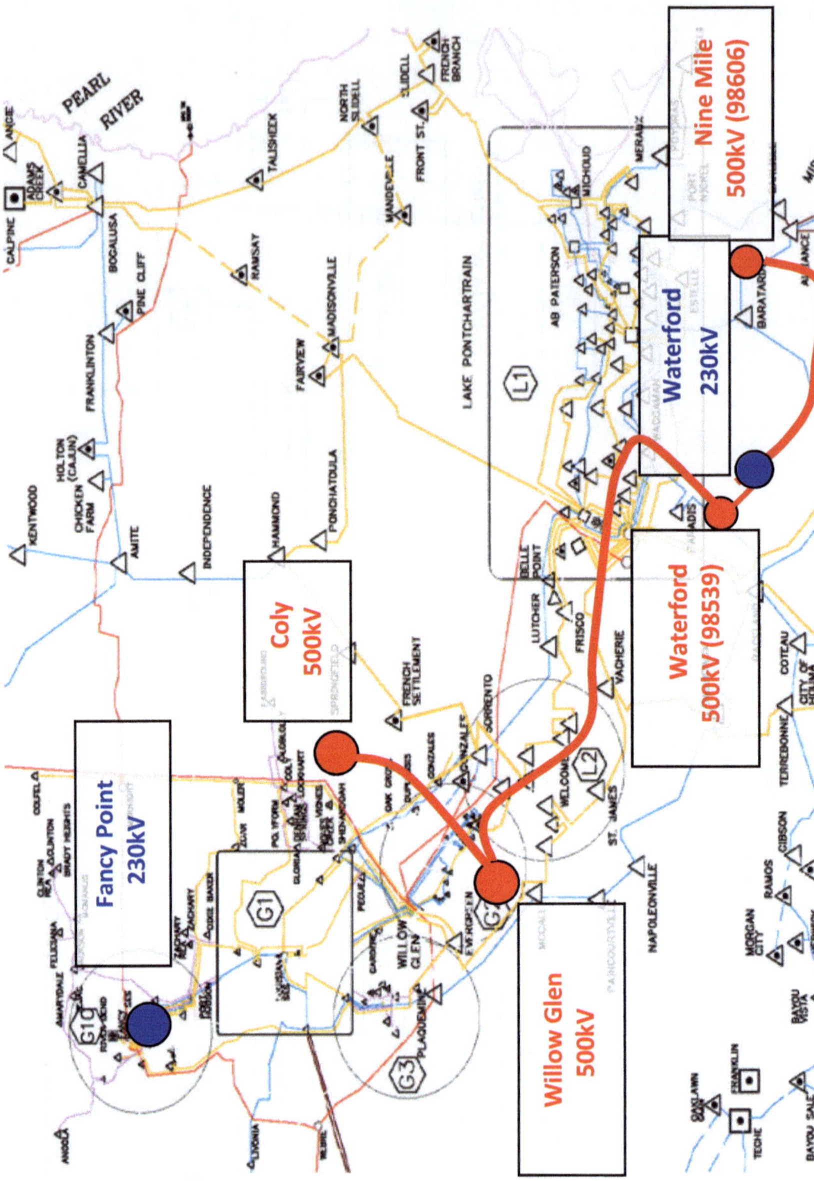

Fig. 4.12 Key transmission path

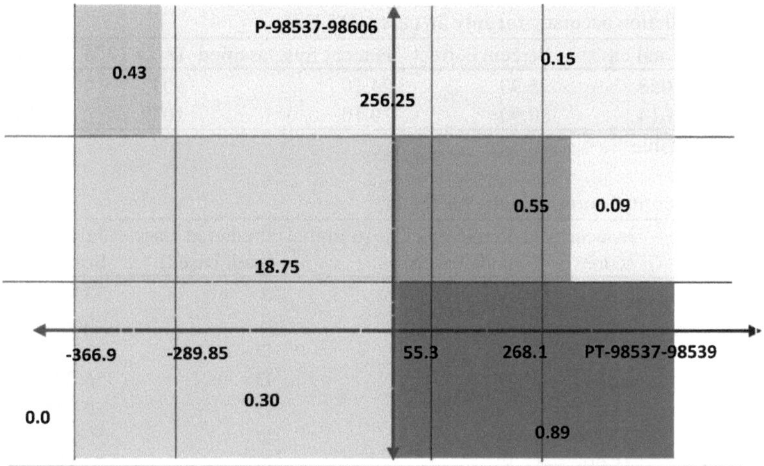

Shade							
Probability of contingency resulting in damping problems	0.00	0.09	0.15	0.30	0.43	0.55	0.89

Fig. 4.13 Operational nomogram built using DT-D2

PT_98537_98539≥55.3 MW and P_98537_98606≤18.75 MW for the current OC, then the chances of any line outage (on 230 and 500 kV lines) resulting in damping problems equals 89 %.

4.14 Evaluation of DT Reliability

To assess the performance of the constructed DTs on unseen cases, further tests were carried out using the data for July 26, 2006 (a week after July 19, 2006). In order to generate different operating scenarios, a procedure similar to the one used for the July 19, 2006 cases was followed. Time domain simulations were carried out for contingencies on all lines above 230 kV. The results were stored in a database. The DT-D2 was tested on this new set of data. The prediction accuracy obtained is as shown in Table 4.13.

Using the new path-based approach proposed in this chapter, the following paths were identified (Table 4.14), and their ISs were calculated. The weight ω_i used for each node is assumed to be equal to the number of learning cases present in that node. Since the importance of a particular node depends on the number of cases in that node, this assumption is valid. All paths with an IS higher than 30 % were considered to be insecure. The rows in gray highlight the terminal nodes whose predicted class is different from that given by the terminal node-based method.

Details regarding the classification results of DT-D2 on the LS cases (July 19) and the test cases (July 26) are given below (Table 4.15). The total number of cases in each terminal node along with the predicted and actual class is tabulated.

Table 4.13 Prediction accuracy for July 26 cases (DT-D2)

Actual class	Total cases	Percent correct	Percent misclassified	D N=1875	S N=4797
D	1058	75.90	24.10	803	255
S	5614	80.90	19.10	1072	4542

D Damping, *S* Stable

Table 4.14 Insecurity scores of paths for DT-D2

Path	Insecurity score	Predicted class (terminal node based)	Predicted class (path based)	July 26, 2006 Secure/insecure
R-1	0.00	S	S	3334/2
R-A-C-2	29.32	S	S	296/22
R-A-B-D-3	39.47	D	D	225/13
R-A-B-D-4	41.69	D	D	56/222
R-A-B-E-5	42.07	S	D	302/38
R-A-C-F-6	33.28	D	D	506/544
R-A-B-E-G-I-7	45.99	S	D	54/193
R-A-B-E-G-I-8	46.58	D	D	8/23
R-A-B-E-G-J-9	44.63	D	D	277/1
R-A-B-E-G-J-10	44.06	S	D	0/0
R-A-C-F-H-L-11	25.50	D	S	0/0
R-A-C-F-H-L-12	24.42	S	S	0/0
R-A-C-F-H-K-M-13	26.80	D	S	0/0
R-A-C-F-H-K-M-14	25.04	S	S	0/0
R-A-C-F-H-K-N-15	26.21	D	S	0/0
R-A-C-F-H-K-N-16	25.03	S	S	556/0

D Damping, *S* Stable

The classification accuracy obtained using the path-based method is shown in Table 4.16. Using the new approach, the accuracy for the insecure cases increases from 75.9–98.72 % and that for the secure cases decreases from 80.90–74.56 %. Since it is more important to ensure that insecure cases are not predicted as secure, the results obtained are promising. The robustness of DT-D2 can be enhanced by using a path-based approach rather than the traditional terminal node-based approach.

4.15 Summary of Results for Entergy System

Several tests were carried out on the Entergy system, and four different DTs were built for transient instability and damping problems using two groups of predictors. For both transient instability and damping problems, the DTs using only PMU measurements as predictors performed as well as DTs, which used several other measurements as candidate attributes. Thus, for the particular operating scenario considered, the PMUs currently installed in the Entergy system seem to be adequate

Table 4.15 Terminal node details for DT-D2

Node	Predicted class	Learning set (July 19, 2006)		Test set (July 26, 2006)	
		Total number of cases in node	S/I (actual class)	Total number of cases in node	S/I (actual class)
1	S	6932	6930/2	3336	3334/2
2	S	253	252/1	318	296/22
3	D	199	64/135	238	225/13
4	D	226	26/200	278	56/222
5	S	552	548/4	340	302/38
6	D	1696	471/1225	1050	506/544
7	S	192	192/0	247	54/193
8	D	26	7/19	31	8/23
9	D	220	126/94	278	277/1
10	S	192	192/0	0	0/0
11	D	34	7/27	0	0/0
12	S	474	474/0	0	0/0
13	D	430	171/259	0	0/0
14	S	359	359/0	0	0/0
15	D	90	55/35	0	0/0
16	S	669	655/14	556	556/0

D Damping, *S* Stable

Table 4.16 Prediction accuracy for July 26 cases using path-based method (DT-D2)

Actual class	Total cases	Percent correct	Percent misclassified	D $N=1875$	S $N=4797$
D	1058	98.72	1.28	1851	24
S	5614	74.56	25.44	1428	4186

D Damping, *S* Stable

for stability assessment. However, this may not be true for other scenarios. The predictors used in the DTs built for damping problems are all from a key transmission path from Coly 500 kV (98390), Willow Glen 500 kV (98246), Waterford 500 kV (98539), Waterford 230 kV (98537), and Nine Mile 230 kV (98606). The power flow in this path is a critical factor that affects damping.

The performance of DT-D2 was tested on the July 26, 2006 cases. The classification accuracy for the cases with damping problems was significantly improved with the path-based approach.

4.16 Voltage Stability Case Study on AEP System

The proposed scheme is tested on an American Electric Power operations snapshot, which serves about 5 million customers and covers 11 states in the USA that are Arkansas, Indiana, Kentucky, Louisiana, Michigan, Ohio, Oklahoma, Tennessee, Texas, Virginia, and West Virginia. This power system contains more than 2400 buses, 100 generators, and 2400 transmission lines with voltage level ranges from 4 to 765 kV. The generation capacity is about 38,000 MW and the total length of transmission network reaches over 39,000 miles. The service areas of the AEP

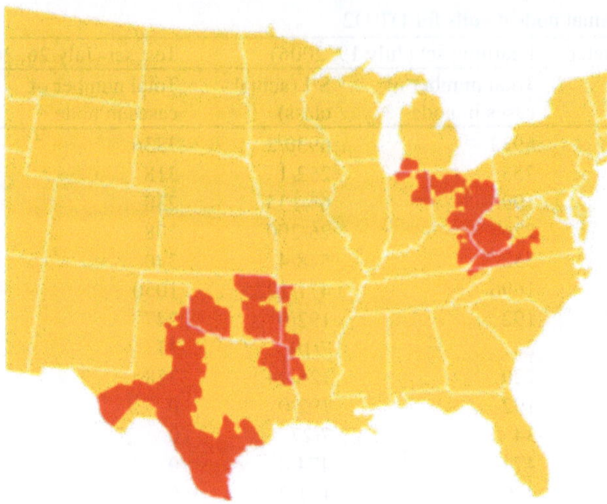

Fig. 4.14 American electric power (AEP) service territories

system are shown in Fig. 4.14. As seen from the figure, the entire AEP system is divided into two regions, the eastern AEP system and the western AEP system, both of which are far away from each other. They are interconnected with the surrounding areas representing the whole eastern interconnected power system in North America. Therefore, a total number of 18,168 buses, 2753 generators, and 19,358 lines are included for voltage security studies.

Currently, there are 27 PMUs located across the eastern AEP system that cover a wide area and their geographical locations are shown in Fig. 4.15. One PMU is installed on a 138 kV bus, 11 PMUs are installed on 345 kV buses while the remaining PMUs are all located on 765 kV buses, which are listed in Table 4.17. These PMUs are installed to monitor the states of the key buses and stations, including bus voltage magnitudes, voltage phase angles, MW and MVAr flows, and current magnitudes of the associated branches. The eastern AEP system is the main focus of this study because the load level is much higher than that of the western AEP system.

In this case study, a software platform involving a variety of simulation tools has been developed to test this scheme. OCs are all generated using the Powerflow & Short-circuit Analysis Tool (PSAT), and voltage security studies are performed using the Voltage Security Assessment Tool (VSAT), both of which are components of the Dynamic Security Analysis Tool (DSA^{Tools}). DSA^{Tools} is an advanced package for power system security evaluation and is developed by the Powertech Labs, Canada [18]. The DTs are trained and tested using a commercial data-mining package, CART, which is developed by Salford System, CA [17]. The database generation, and data conversion and analysis work are conducted using MATLAB and Microsoft Visual Studio VC++ codes.

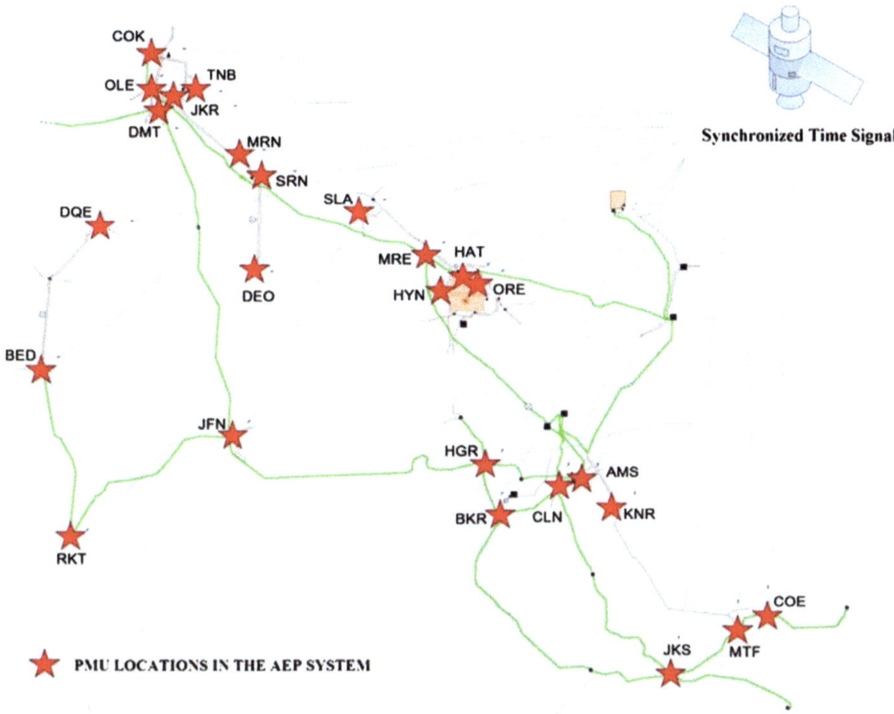

Fig. 4.15 PMU locations in the eastern AEP system

4.17 OC Generation

In this step, 29 OCs are generated in PSAT based on the generation and load patterns provided by the AEP operations staff. They represent stressed OCs that include all the details of load levels, generator outputs and branch power flows during a specific period of time. Some of these OCs stress the system well beyond the normal operating ranges. The load levels of these OCs are depicted in Fig. 4.16. The voltages of all the buses in the AEP system are within reasonable levels at the base case:

1. All the bus voltage magnitudes are adjusted to be between 0.90 and 1.10 in p.u.
2. The voltages of 138–765 kV buses are adjusted in the range of 0.95–1.10 p.u.

Table 4.17 PMU locations in the eastern AEP System

No.	Bus name	Bus no.
1	Rockport 765 kV	4974
2	Kanawha River 345 kV	195
3	Jacksons Ferry 765 kV	1409
4	Matt Funk	1440
5	Orange 765 kV	2221
6	Cook 345 kV	5599
7	Olive 345 kV	5872
8	Dumont 345 kV	5648
9	Dumont 765 kV	5649
10	Dequine 345 kV	4808
11	Twin Branch 345 kV	5983
12	Jackson Road 345 kV	5746
13	Meridian 345 kV	4913
14	Sorenson 345 kV	5407
15	Desoto 345 kV	4811
16	Breed 345 kV	4767
17	Jefferson 765 kV	4872
18	Sw Lima 345 kV	4186
19	Marysville 345	3900
20	Marysville 765 kV	3901
21	Heyden 345 kV	4491
22	Haytt 345 kV	4513
23	Hanging Rock 765 kV	2116
24	Baker 765 kV	539
25	Culloden 765 kV	129
26	Amos 765 kV	35
27	Cloverdale 765 kV	1277

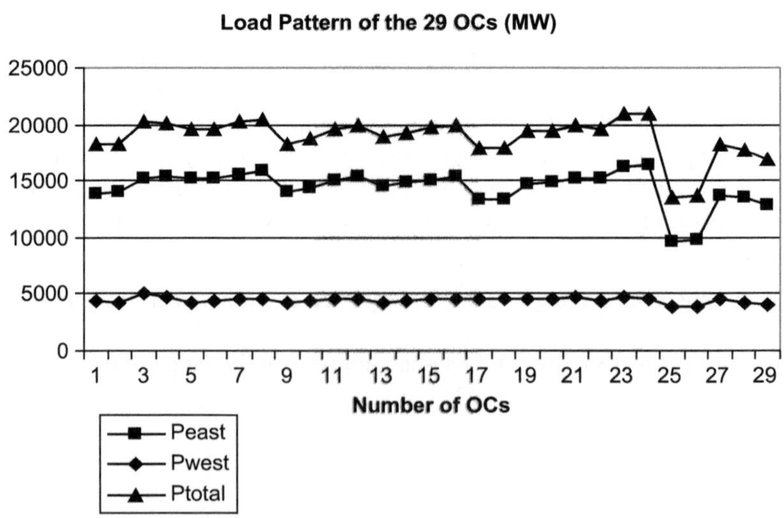

Fig. 4.16 Load pattern on the 29 stressed OCs

4.18 Voltage Security Analysis

A list of critical contingencies that may cause severe voltage problems is selected according to previous operating experiences in the AEP system. This list contains 163 critical contingencies including different transmission line and transformer outages. This research project adopts a static analysis method to evaluate postcontingency voltage security and a voltage collapse judgment is given when a contingency results in the divergence of the power flow solution [1, 19]. Thus, the database consists of $29 \times 163 = 4727$ voltage security simulations with either security or insecurity labels marked on each case.

4.19 Predictor Selection and Database Generation

Since all the voltage insecure cases are caused by contingencies in the list, the voltage security predictors for DT training should include the contingency-dependent information together with contingency-independent parameters in the system. For the contingency-dependent parameters, an unordered bus pair x and y is adopted to denote the faulted transmission line or transformer, because the voltage security analysis method estimates voltage security by solving power flows without the faulted branch, and moreover, it does not need any detail of the fault such as type, duration, or location. The only useful information is the bus locations of the faulted branch. To eliminate the ordering of the two buses of a critical branch during DT training, each contingency case is doubled in the form of "bus-1 $=x$ and bus-2 $=y$" and "bus-1 $=y$ and bus-2 $=x$," which treats the two buses equally. As a result, a database containing $4727 \times 2 = 9454$ cases is generated. This measure allows the DTs to identify buses or substations common to contingencies that are more prone to cause voltage collapse. The contingency-independent parameters reflect critical system states in real time; therefore, they are chosen from the PMU-monitored predisturbance system parameters in the AEP system. In order to compare the performance of different parameters in capturing postcontingency voltage behaviors, eight groups of predictors are used as defined in Table 4.18.

Group 1 uses a categorical bus pair (FB and OB) to represent the contingency branch. Group 2 selects voltage phase angle differences among all of the 27 existing PMUs (A_x_y), where x and y are the PMU bus numbers. There are $27 \times 26/2 = 351$ angle differences chosen as voltage security predictors. Although the information contained in the 351 angle differences are redundant since only 27 values are collected, the DT only picks up the most effective ones as CSRs. The predictors in groups 1 and 2 help to compare the DT performance between contingency-dependent and contingency-independent parameters. Group 3 evaluates the DT performance using the combination of these two types of predictors. Groups 4 and 5 replace the phase angle differences with all the available current magnitudes and reactive power magnitudes on branches measurable by the PMUs. Groups 6 and 7 test the performance

Table 4.18 Different PMU-related predictor groups in the AEP system

No.	Predictors
Group 1	FB and OB of the contingency branch
Group 2	Voltage phase angle differences (*A_x_y*)
Group 3	*A_x_y*, FB, OB
Group 4	Current magnitudes on branches (*I_x_y*), FB, OB
Group 5	MVAr flows on branches (*Q_x_y*), FB, OB
Group 6	Square of voltage magnitudes (V^2_*x*), FB and OB
Group 7	Absolute value of current magnitude multiplied by branch impedance (*IZ_x_y*), FB and OB
Group 8	*A_x_y*, *Q_x_y*, *I_x_y*, V^2_*x*, *IZ_x_y*, FB, OB

FB faulted bus, *OB* other bus

Table 4.19 Sample of the created databases

SECU$	FB$	OB$	*Q*_35_129	*I*_4,513_4515	*A*_35_129	…
S	1998	2266	−398.5	13.4	0.9	
S	2266	1998	−398.5	13.4	0.9	
I	5983	9319	−398.5	13.4	0.9	
I	9319	5983	−398.5	13.4	0.9	*for OC1*
S	1205	1207	−398.5	13.4	0.9	
S	1207	1205	−398.5	13.4	0.9	
…	…	…	…	…	…	
I	1998	2266	−410	15	1.2	…
I	2266	1998	−410	15	1.2	
S	5983	9319	−410	15	1.2	
I	9319	5983	−410	15	1.2	*for OC2*
S	1205	1207	−410	15	1.2	
S	1207	1205	−410	15	1.2	
…	…	…	…	…	…	…

SECU Voltage Security Label, *FB* faulted bus, *OB* other bus, *S* secure, *I* insecure

of square of voltage magnitudes on PMU buses and the absolute value of the current magnitude multiplied by the branch impedance. The selection of these contingency-independent predictors is designed to collect critical system information that can indicate voltage problems more precisely and intuitively. *A_x_y* and *I_x_y* are good measures to indicate the degree of stress at an OC; *Q_x_y* plays a more important role than active power flow (*P_x_y*) in supporting voltage profiles; V^2_*x* is more sensitive than voltage magnitude itself especially after contingencies; and *IZ_x_y* covers more information than current magnitude in reflecting voltage problems. Finally, group 8 includes all the predictors indicated above in order to compare their performance in the DTs.

With all the case labels and predictors collected, eight databases are generated in MATLAB codes for the eight groups of predictors, respectively. One sample of the created databases is shown in Table 4.19. The first column is the voltage security label obtained from VSAT results (S or I), the second and third columns represent the FB and the other bus (OB), respectively, and the remaining columns are differ-

Table 4.20 Performance comparison of the optimal DTs

Opt. DTs	Size	Leaning set accuracy (%)			Test set accuracy (%)		
		I	S	Overall	I	S	Overall
DT1	24	71.4	63.31	66.08	68.82	60.95	63.72
DT2	7	87.61	76.99	80.63	86.96	80.07	82.50
DT3	36	98.69	94.33	95.82	91.00	93.63	92.70
DT4	50	99.50	95.66	96.97	92.95	94.93	94.24
DT5	31	98.11	93.89	95.33	87.11	93.3	91.12
DT6	38	96.49	92.42	93.81	88.91	91.91	90.85
DT7	55	99.69	95.19	96.73	91.00	93.46	92.6
DT8	51	99.11	96.00	97.06	90.55	94.44	93.07

DT decision tree, *I* insecure, *S* secure

ent types of measured values from PMUs. The "$" sign in the first three columns is used to differentiate the categorical variables from the numerical ones. Each row in the database represents one voltage security simulation case following one critical contingency. The predisturbance system parameters for each OC are the same from case to case, and the only difference rests with the contingency bus pairs, FB and OB. From the simulation results in VSAT, the database contains 3258 cases out of 9454 cases (34.46 %) that fail to obtain a convergent power flow solution following a single branch outage. They are regarded as "I" cases leading to voltage collapses. The remaining cases are labeled as "S" cases.

4.20 DT Training and Performance

All the 9454 cases in the created database are given equal weight and 1891 cases (20 %) are randomly selected to form a TS. The remaining 7563 ones (80 %) are used to form the LS. In order that fewer insecure cases are misclassified as secure, the cost of misclassifying an insecure case to be a secure case is reasonably increased. Also, different algorithms including Gini, symmetric Gini, entropy, class probability, twoing, and ordered twoing are tested and compared during the DT training progress to obtain the optimal DT [20]. Therefore, eight optimal DTs from DT1 to DT8 are trained for the above databases, respectively, and the prediction accuracy for both LS and TS are all listed in Table 4.20.

From Table 4.20, a comparison of DT1 and DT2 indicates that contingency-independent parameters are more important than contingency-dependent parameters in voltage security prediction at stressed OCs; this is because DT2 performs much better than DT1 for both secure and insecure cases. One explanation is that the system OC is more critical than fault location in voltage collapse prediction. Voltage phase angle differences can effectively indicate the degree of stress at an OC and large values of *A_x_y* usually indicate a more stressed OC. A severely stressed OC is more vulnerable to voltage collapse than a lightly loaded one following the same contingency. Similar tests were conducted by using the other contingency-independent parameters alone as predictors such as *I_x_y*, *Q_x_y*, V^2*_x*, and *IZ_x_y*,

Table 4.21 Number of critical system attributes selected in the optimal DTs

DTs	*Ns/Nt*	DTs	*Ns/Nt*
DT3	15/353	DT6	13/29
DT4	16/87	DT7	17/87
DT5	12/87	DT8	18/635

DT decision tree

but improvement in prediction accuracy was not observed. Therefore, the remaining predictor groups in Table 4.18 (group 3–8) also include the contingency-dependent information for DT performance improvement, and the results in DT3–DT8 have proved this. The combination of FB, OB, and various contingency-independent parameters is helpful for building much better DTs compared to DT2.

In DT3, the prediction accuracy for the TS is increased to 91.0 % for insecure cases and 93.63 % for secure cases. The overall accuracy is 92.7 %. A better DT (DT4) with higher accuracy for the TS is obtained by using current magnitudes as predictors, which can accurately predict 92.95 % insecure cases and 94.93 % secure cases. Similarly, DT5, DT6, and DT7 are trained, respectively, by including values of *Q_x_y*, V^2*_x*, and *IZ_x_y* from the same PMU locations, and their performance is similar to DT3. DT8 attempts to cover all the above predictors for accuracy improvement in the TS because it includes the most system information. Unfortunately, the overall accuracy for the TS in DT8 does not see any significant improvement. Instead, the overall accuracy drops to 93.07 %, which is even lower than DT4, although it has a better performance on the LS. This is mainly caused by the over-fitting problem in the DT training process, which uses too many parameters as predictors to build one DT. Another problem occurs because some of the important parameters are masked behind the CSRs in the DT since each splitting of the internal nodes guarantees the highest impurity reduction for the current node, instead of optimizing the performance of the whole tree.

To illustrate the efficiency of the DTs that are trained, Table 4.21 lists the total number of CSRs selected in DT3–DT8. *Ns* is the number of CSRs in the DT, while *Nt* represents the total number of predictors in the corresponding databases. It can be observed that only a small portion of the predictors are selected to be CSRs, indicating that the voltage security assessment process in real time is very fast. The important system parameters can be used to build an online nomogram in identifying voltage security boundary and also to provide an effective measure for the operators to steer the system conditions away from the security limits after severe disturbances occur.

4.21 DT Performance Improvement

The results in Sect. 4.20 indicate that the best performance on the TS using a single DT is in DT4, the overall accuracy of which is 94.24 %. In other words, there are 5.76 % cases misclassified using this tree model. The misclassification problem in

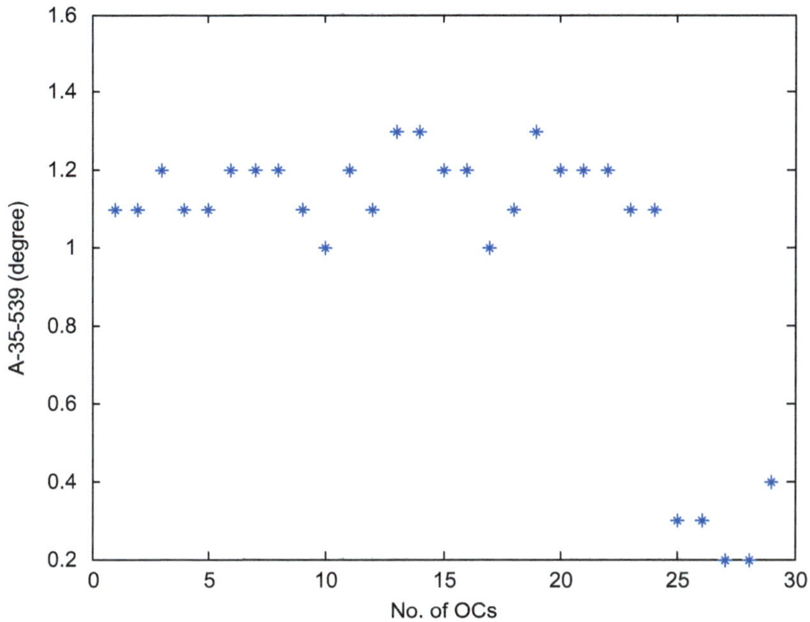

Fig. 4.17 Distribution of *A*_35_539 for the original 29 OCs

DTs could be caused by a variety of reasons, among which the formation of the LS and TS plays an important role. If the cases in the TS have great similarities with the ones in the LS, the DT has better performance on the TS. On the contrary, high accuracy of prediction on totally unknown cases may not be guaranteed. By periodically including new and unknown system states into the database, DTs are updated to learn more useful information for improving robustness and the classification accuracy can be effectively increased.

Another observation is that DT performance in predicting voltage security depends on the distribution of the CSRs and a preliminary test is used to illustrate this phenomenon. Focusing on one CSR that has a severe misclassification problem in a DT trained by using *A*_35_539, FB, and OB, it is found that most of the values of this parameter in the 29 OCs are concentrated in a small range, as shown in Fig. 4.17. The horizontal axis represents the number of different OCs, while the vertical axis stands for the phase angle difference between bus 35 and bus 539 (in degrees). The values of *A*_35_539 range from 0.2 to 1.3 and most of them are concentrated in the range from 1.0 to 1.3. Only five OCs cover the range from 0.2 to 0.4. The values from 0.4 to 1.0 are totally missing. The distribution of these values shows a severe imbalance, which could be an important factor causing the misclassifications. As a result, more OCs are created for the values of *A*_35_539 to be distributed more evenly in this range. To accomplish this, generator active power outputs around bus 35 and bus 539 and the load levels in the AEP system are adjusted. This can effec-

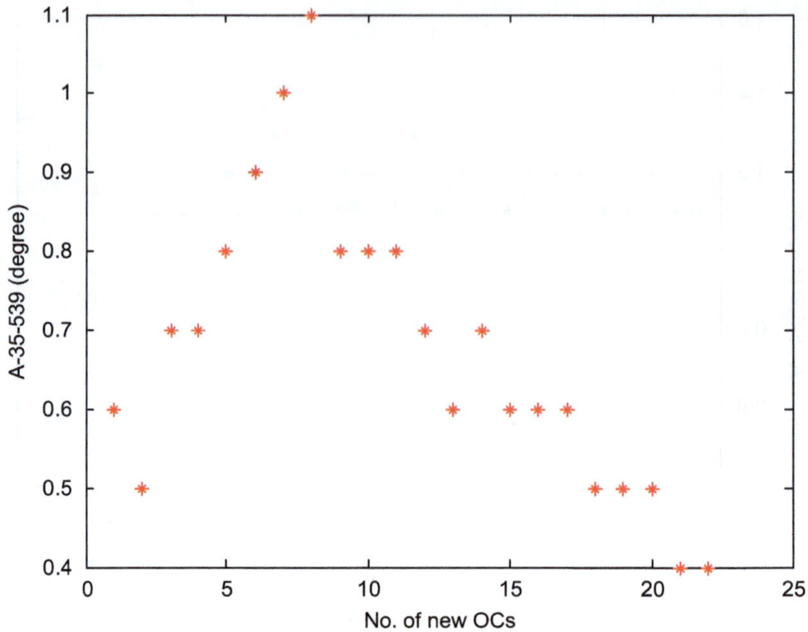

Fig. 4.18 Distribution of *A*_35_539 for the new 22 OCs

tively change the values of *A*_35_539 in the desired pattern. Twenty-two (22) new OCs are then created with the new distribution of *A*_35_539 shown in Fig. 4.18. Another database, including $22 \times 163 \times 2 = 7172$ cases, is created for the DT training, in which there are 288 insecure cases (4%). We can also observe that *A*_35_539 becomes a very important CSR in the new DT, which is located in the second level. This test indicates that by adjusting the previously problematic variable to cover more information near the splitting boundary, it can yield a better DT.

However, the proposed scheme for online voltage security assessment should work well for all kinds of unforeseen OCs no matter how the critical system parameters are distributed. Therefore, several methods to improve DT performance for online applications are discussed below when the future OCs for the next hour are relatively fixed.

4.21.1 Multiple Optimal DTs

As discussed in Sect. 4.20, the combination of all the available predictors may not improve the DT performance on the TS because of the over-fitting and variable masking problems. In order to improve reliability and prediction accuracy, an approach using multiple optimal DTs that are trained by combinations of different predictors is presented. This method takes advantage of all the DTs that satisfy a de-

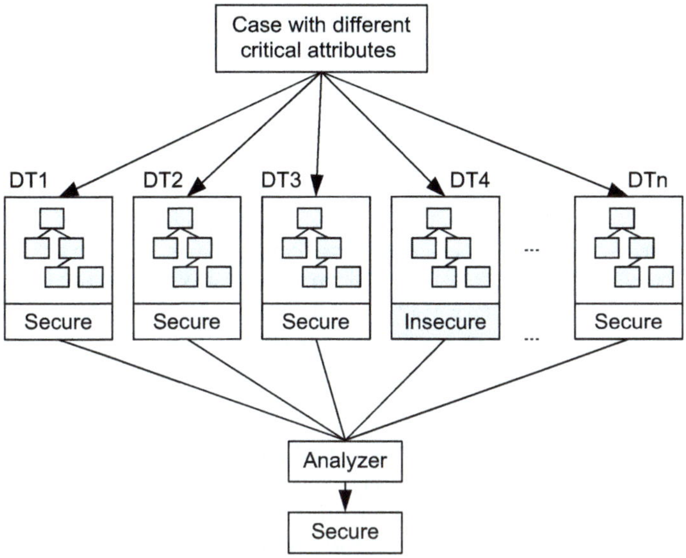

Fig. 4.19 Concept of multiple DT application

sired threshold of performance instead of trusting the tree with the best performance only. The cases that are misclassified by the best DT may be correctly predicted by the other trees that use different PMU-related CAs. For online applications, all the optimal DTs are used to obtain a comprehensive classification result and the basic flow chart of this idea is shown in Fig. 4.19. For any incoming case, using multiple DTs can obtain different prediction results. As an example, although DT4 provides an insecure result, all the other DTs provide secure results. Therefore, a comprehensive result, "secure," is provided, which is considered as the final prediction result.

Based on the predictor groups introduced in Sect. 4.19, a heuristic search is conducted to identify different combinations of these predictors that contribute to good DTs. The multiple DTs that are used for online application should be sufficiently different from each other because certain types of predictors can be totally masked when combined with other predictors to build a DT resulting in exactly the same tree as the one trained before predictor combination. For example, when the predictor V^2_x was combined with any other type of predictors, they were never selected as CSRs. Thus, all different combinations of A_x_y, Q_x_y, I_x_y, and IZ_x_y are tested to create optimal DTs with sufficient difference. Nine (9) DTs with good performance on the TS are obtained and shown in Table 4.22.

The DTs from DT9 to DT17 are all different from each other although some of the branches share the CSRs. The result shows most of them offer better prediction performance on the TS than the DTs trained before predictor combination. As a result, all the 15 DTs from DT3 to DT17 are used for online application since they are capable of collectively predicting over 91 % of cases in the TS. For any contingency case, 15 voltage security assessments are obtained separately and a final assess-

Table 4.22 DT performance for different predictor combinations

Opt. DTs	Size	Predictor combinations	Test set accuracy (%)		
			I	S	Overall
DT9	59	*A_x_y*, *I_x_y*, FB, OB	90.4	95.59	93.76
DT10	49	*A_x_y*, *Q_x_y*, FB, OB	89.21	95.02	92.97
DT11	47	A_x_y, IZ_x_y, FB, OB	90.85	94.12	92.97
DT12	55	*I_x_y*, *Q_x_y*, FB, OB	92.95	95.26	94.45
DT13	45	Q_x_y, IZ_x_y, FB, OB	91.00	93.79	92.81
DT14	53	A_x_y, I_x_y, Q_x_y, FB, OB	92.95	95.18	94.39
DT15	47	A_x_y, I_x_y, IZ_x_y, FB, OB	90.85	94.12	92.97
DT16	53	I_x_y, Q_x_y, IZ_x_y, FB, OB	91.15	93.14	92.44
DT17	45	A_x_y, Q_x_y, IZ_x_y, FB, OB	91.00	93.79	92.81

DT decision tree, *FB* fault bus, *OB* other bus, *I* insecure, *S* secure

ment is given in terms of the majority classification results. A statistical analysis is conducted on 1891 cases in the TS and 7563 cases in the LS using these 15 DT models. The results indicate that there are only 101 cases and 179 cases that are misclassified eight times or more in the TS and LS, respectively. Therefore, the overall prediction accuracy on the TS is increased to 94.66 %, while the accuracy on the LS is increased to 97.75 %. Although the overall improvement is limited, the results obtained from a variety of DTs using different CAs are more convincing than that of only one tree. In addition, using multiple DTs in real time will not cause a significant increase in the total security assessment time because the DTs are all trained offline and the assessment process using different DTs can be conducted individually.

4.21.2 Corrective DTs

The above method is designed to build many DTs by using different CAs for successful classifications on the misclassified cases in the TS. The main problem is that the absolute computation burden is increased although the DT training process can be conducted in parallel. According to the observation that most of the paths in a single DT trained by only one type of contingency-independent parameters (e.g., DT3) have excellent prediction behaviors on the TS and only a few paths have severe misclassification problems, another idea is developed to partially modify the DT by replacing the problematic paths with corrective DTs for accuracy improvement. For each of the paths with poor performance, a corrective DT is trained by including more system information for all the cases that fall into this path in the original database. During the voltage security assessment process, these problematic branches in the original DT are not abandoned; instead they are linked to these corrective DTs for further classification. For example, a path with poor performance in DT3 is taken to explain how this method works and this path is plotted in Fig. 4.20a.

DT3 is trained using the LS containing 7563 cases, among which there are 49 (24 insecure and 25 secure) cases falling into Terminal Node 4. Fourteen (14) out of 1891 cases in the TS are tested using this path and 6 secure cases are misclassi-

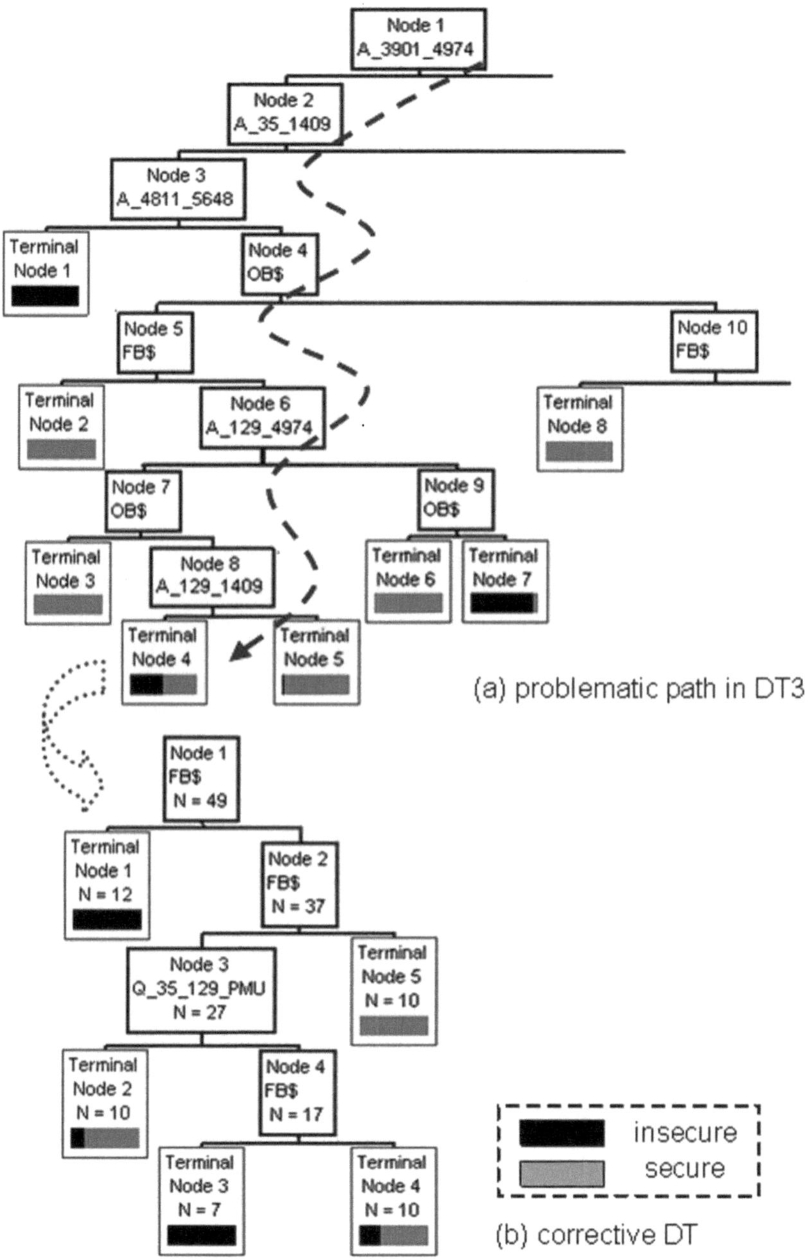

Fig. 4.20 Corrective DT for a problematic path in DT3

Table 4.23 Performance of the corrected DTs

Opt. DTs	Overall LS accuracy (%)		Overall TS accuracy (%)	
	Corrected	Original	Corrected	Original
DT3'	97.46	95.82	94.02	92.70
DT4'	98.21	96.97	*95.45*	94.24
DT5'	98.70	95.33	92.07	91.12
DT6'	97.38	93.81	93.39	90.85
DT7'	98.40	96.73	93.92	92.6

DT decision tree, *LS* learning set, *TS* test set

fied as insecure cases. Although higher accuracy for these 49 cases in the LS can be achieved by further splitting Terminal node 4, it will jeopardize the prediction performance on the TS. The problem lies in using *A_x_y* alone, as this information is not enough to clearly classify these cases. Therefore, the 49 cases in the LS and the 14 cases in the TS are picked to form a smaller learning set (LS') and a test set (TS'), respectively, but more predictors like *I_x_y* and *Q_x_y* are included. A corrective DT using LS' and TS' is now trained and it only misclassifies five cases in the LS' and three cases in the TS'. This small DT (depicted in Fig. 4.20b) performs much better than the original path that reaches Terminal Node 4 in DT3 for these 49 + 14 = 63 cases. In real-time application, the original DT3 is first used to investigate voltage security. The cases that fall into Terminal Node 4 are further evaluated using this corrective DT to obtain the final security assessment. This approach preserves the branches with high accuracy and only modifies the paths with severe misclassification problems by using corrective DTs rather than re-training the whole tree. This method is tested on all the DTs from DT3 to DT7 that are trained using only one type of contingency-independent predictors; the improved DT performance is shown in Table 4.23.

From Table 4.23, all of the original DTs show improvement, with better performance for both LS and TS. The best performance is achieved in DT4' and it can correctively predict 95.45 % cases in the TS, which has a 1.21 % improvement. During the training of corrective DTs, not all of the problematic paths can be improved by adding more predictors for further classification. The main reasons are: (a) there are not enough cases falling into the problematic path to train a corrective DT and (b) the cases in the LS that are classified by the problematic path all belong to one class, secure or insecure, which makes it impossible to build the corrective DT. Further investigation of this "multi-level DT" idea to improve prediction accuracy will be carried out in future work.

4.21.3 Maximum DTs

The third measure for increasing DT accuracy is by using the maximum trees with high accuracy for the whole database. Critical system states for the next hour can be obtained an hour ahead with the help of load forecasting techniques and timely

Table 4.24 Maximum DT performance on the whole database

Max. DTs	Size	Accuracy for the whole database (%)		
		I	S	Overall
DT3_M	117	99.91	98.72	99.13
DT4_M	125	99.85	98.68	99.08
DT5_M	112	99.85	99.11	99.37
DT6_M	142	99.97	98.66	99.11
DT7_M	126	99.85	99.05	99.32
DT8_M	125	99.91	99.11	99.39

DT decision tree, *I* insecure, *S* secure

updated system information. In this situation, almost all the incoming OCs for the next hour with sufficient accuracy can be included in the database, and the DT performance on the whole database becomes more important to system operators. Therefore, the trained maximum DTs with high accuracy on the whole database are a good choice for online applications in the industry.

To obtain a more reliable DT model, the "V-fold Cross Validation" method described in [20] is used to form the LS and TS. This method divides the whole database into n groups with equal number of cases, uses each group once as the TS while the remaining $n-1$ groups form the LS, and finally obtains a comprehensive DT based on the n generated DTs. Several tests were conducted to train the maximum DTs for the databases created in Section IV_C by setting $n=10$. The results in Table 4.24 indicate that any single DT trained from predictor group 3–8 has an overall accuracy higher than 99.08 %. More importantly, the overall prediction behaviors of these maximum DTs on the TS are around the level of 90 %, which indicates that using the maximum trees for online application does not lose the prediction ability on the unknown cases. Higher accuracy on the whole database can be further achieved by setting the whole database as one LS without any TS, but the corresponding maximum DT may have poor performance when system condition changes. The main problem of this maximum tree method is the high degree of complexity; however, there is no need to install new PMU devices because all the predictors are collected from the existing PMUs.

References

1. P. Kundur, *Power System Stability and Control* (McGraw-Hill, New York, 1994)
2. M. Pavella, D. Ernst, D. Ruiz-Vega, *Transient Stability of Power Systems: A Unified Approach to Assessment and Control* (Kluwer, Norwell, 2000)
3. L. Wehenkel, *Automatic Learning Techniques in Power Systems* (Kluwer, Norwell, 1998)
4. Houben, L. Wehenkel, M. Pavella, Coupling of K-NN with decision trees for power system transient stability assessment. *Proceeding of 4th IEEE conference on control applications*, pp. 825–832, Sept 1995
5. L. Wehenkel, T. Van Cutsem, M. Pavella, An artificial intelligence framework for online transient stability assessment of power systems. IEEE Trans. Power Syst. **4**(2), 789–800 (1989)

6. A. Fouad, T. Jianzhong, Stability constrained optimal rescheduling of generation. IEEE Trans. Power Syst. **8**(1), 105–112 (1993)
7. E.S. Karapidakis, N.D. Hatziargyriou, Online preventive dynamic security of isolated power systems using decision trees. IEEE Trans. Power Syst. **17**(2), 297 − 204 (2002)
8. L. Wehenkel, T. Van Cutsem, M. Pavella, Inductive inference applied to on-line transient stability assessment of electric power systems. Automatica **25**, 445–451 (1989)
9. L. Wehenkel, M. Pavella, E. Euxibie, B. Heilbronn, Decision tree based transient stability method a case study. IEEE Trans. Power Syst. **9**(1), 459–469 (1994)
10. L. Wehenkel, M. Pavella, Decision trees and transient stability of electric power systems. Automatica **27**(1), 115–134 (1991)
11. L. Wehenkel, M. Pavella, Advances in decision trees applied to power system security assessment. *Proceeding of IEE 2nd international conference on advances in power system control, operation and management*, pp. 47–53, Dec 1993
12. S. Rovnyak, S. Kretsinger, J. Thorp, D. Brown, Decision trees for real-time transient stability prediction. IEEE Trans. Power Syst. **9**(3), 1417–1426 (1994)
13. K. Mei, S. Rovnyak, Response-based decision trees to trigger one-shot stabilizing control. IEEE Trans. Power Syst. **19**(1), 531–537 (2004)
14. N. Senroy, G.T. Heydt, V. Vittal, Decision tree assisted controlled islanding. IEEE Trans. Power Syst. **21**(4), 1790–1797 (2006)
15. L. Breiman, J. Friedman, R.A. Olshen, C.J. Stone, *Classification and Regression Trees* (Wadsworth Statistics/Probability) (Chapman and Hall/CRC, London, 1984)
16. TSAT Transient Security Assessment Tool. http://www.dsapowertools.com/downloads/TSAT_Brochure.pdf
17. CART®Tree-Structured Non-Parametric Data Analysis. http://www.salfordsystems.com/cart.php
18. DSATools Dynamic Security Assessment Software. http://www.dsatools.com/
19. T. Van Cutsem, C. Vournas, *Voltage Stability of Electric Power Systems* (Kluwer, Norwell, 1998)
20. Salford System, CART for windows user's guide-an implementation of the original CART methodology. Software Manual, 2002

Chapter 5
Transmission Line Fault Location

5.1 Fault Location Applications

Fault location on transmission lines is an important application in power systems [1]. It allows operators, as well as protection and maintenance staff, to pinpoint where the fault has occurred and initiate corresponding effort to repair and restore the affected transmission line. Once the protective relays have detected a transmission line fault and tripped (disconnected) the line by issuing a control command to circuit breakers, the fault location application determines the location by measuring voltages and currents at either or both transmission line ends. Over the years, many different fault location techniques were developed, but very few have used synchronized samples or synchrophasors [2]. Most recently, the use of synchronized samples and phasors in fault location has become a subject of interest since the technology has matured and is readily available [3].

The synchronized samples or synchrophasors provided by phasor measurement units (PMUs) or other intelligent electronic devices (IEDs) may be utilized in fault location applications in the transmission system in many different ways. Two techniques will be discussed as follows:

1. Using synchronized samples taken at two ends of a transmission line
2. Using sparse measurements of synchronized phasors across a region

The most accurate fault location is feasible with the method that uses synchronized samples taken at both ends of a transmission line. If the circumstances in the power system, such as tapped lines and lines that do not have synchronized measurements available at each end are present, then a technique that uses sparse measurements may be used; this technique is less accurate but still much better than any other manual techniques available. Synchronization of samples or phasors is made available through the use of GPS receivers, and hence fault location improvements are dependent on the availability of such receivers [4]. Many utilities today are considering the use of precise time clocks that are provided through the GPS system, so it is expected that fault location that requires use of the precise clock will be readily available and widely used in the future. The applications that are described next

M. Kezunovic et al., *Application of Time-Synchronized Measurements in Power System Transmission Networks,* Power Electronics and Power Systems,
DOI 10.1007/978-3-319-06218-1_5,

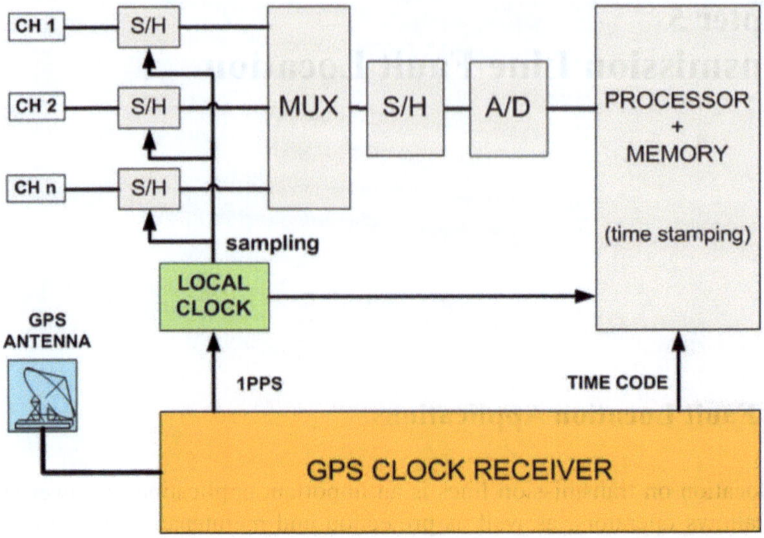

Fig. 5.1 Internal design of a recording device that provides synchronous sampling. *MUX* multiplexer

have been studied extensively, and ample test results, including using field data, are used to evaluate the feasibility and practical benefits.

5.2 Implementation Issues

To allow implementation of the proposed fault location techniques, a particular internal design of the waveform sampling interface is required. This design is used to provide sampling synchronization features, as shown in Fig. 5.1.

Figure 5.1 illustrates the following: the sampling device has to have a connection between its internal clock, which strobes the sample and hold (S/H) circuits for the sampling to take place, and the external clock signal, which in turn is offered as a pulse-per-second (PPS) output of a GPS receiver. In addition, such a device must be capable of utilizing the time code from the GPS receiver so that it can time-stamp the set of samples or calculated values once they are converted into a digital format (computer word) by the analog-to-digital (A/D) converter. The multiplexer (MUX) allows independent conversion of each signal (channel) by switching one phase of the analog signal to the next, as the conversion is initiated/completed. In reality, the interfacing to a GPS receiver may have many variations since the PPS and time code may be taken from a local GPS receiver or obtained from a remote one through communication means. Such details will be shown later when a variety of substation connections is discussed.

For the proposed techniques to be implemented, an IED with the mentioned design features needs to be available in respective substations.

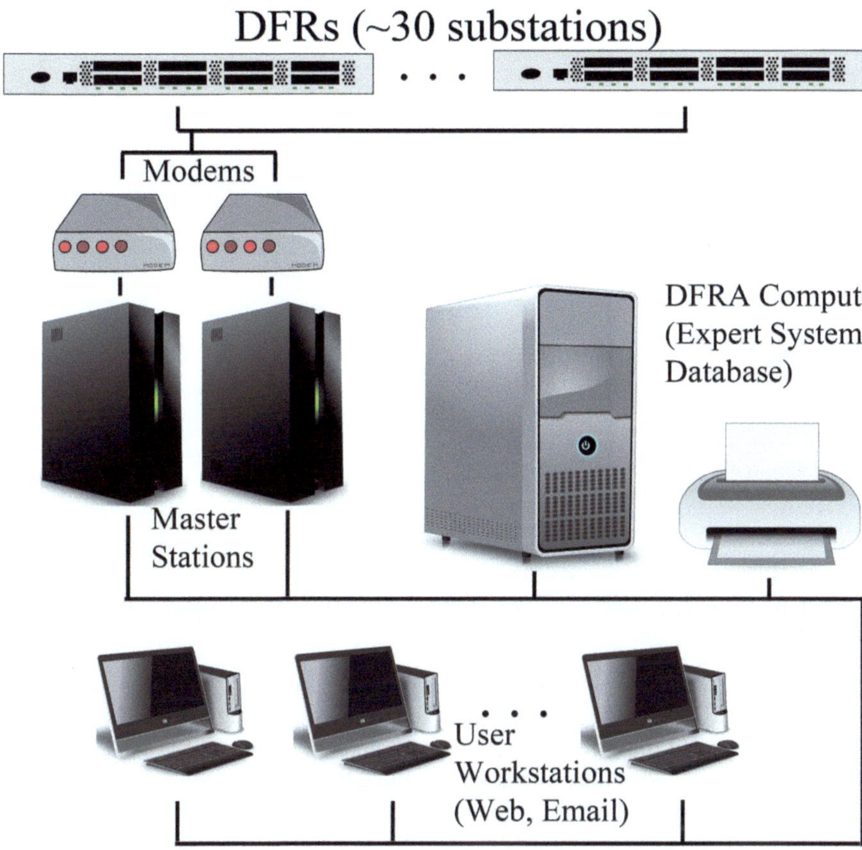

Fig. 5.2 Typical remote access connection of digital fault recorders (DFR)

Digital fault recorder (DFR) This type of recording device samples currents and voltages across a given substation with sampling rates of several kilohertz. The main feature of such a device is the ability to get triggered by a disturbance, and then to record the transient waveforms for an extended period of time covering the disturbance of interest. Most of the DFRs in service today have a remote access capability where the recorded samples can be uploaded to a master station located in a remote site such as a control center of protection staff offices. Once connected to a GPS of satellites through a GPS receiver, the samples can be synchronized across a transmission line end. Several vendors offer PMU functionality as an integral part of the DFR design. Figure 5.2 shows a connection of several DFRs to a centralized system where multiple master stations, data analytics server, and user workstations are interconnected for collecting, further processing, and viewing DFR data, respectively.

Phasor Measurement Unit (PMU) PMUs are installed with the GPS receiver being an integral part of the basic function, which is to measure phasors synchronously

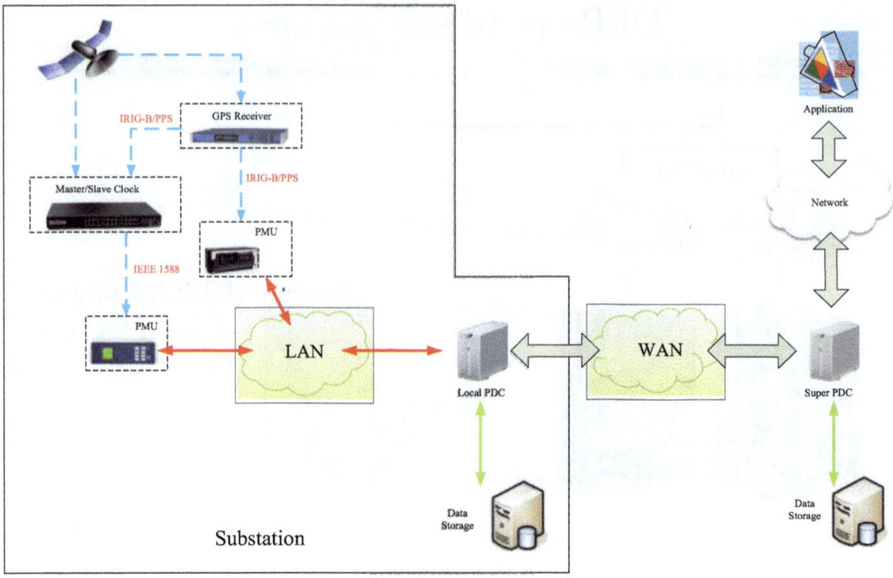

Fig. 5.3 Typical remote access connection of PMU. *PMU* phasor measurement unit, *PPS* pulse per second

over a wide area. The PMU design allows for a high sampling rate in the range of tens of kilohertz, and the samples are used to calculate phasors of voltage and current, which are reported at a rate between 30 and 240 times a second. While it is not a common feature today, PMUs can be easily designed to deliver both samples and phasors as needed. In current practice, only the calculated phasors are uploaded to a remote site where phasor data concentrators (PDCs) are used to manage further transfer or store the phasors. Figure 5.3 shows how PMUs may be connected in a system using different type of GPS interfaces.

Digital Protective Relay (DPR) Protective relays are used to sense faults at each end of a transmission line and trip the breakers once the fault is detected. As a part of its design, a DPR may have several data export functions to allow uses of data for applications other than protection. One such data export function is to capture and upload the waveform samples. This application is very similar to the recording feature of the DFR except the samples are synchronized only across a given transmission line. If a relay is connected to a GPS receiver, then the samples may be synchronized across two ends of a transmission line or even across several relays in a given region. An example of such connection is shown in Fig. 5.4. Several DPRs on the market today also provide a PMU function as an integral part of the basic product functionality. Many vendors provide a file upload management infrastructure with products supporting synchrophasor extraction, communication, and storage at a remote site.

Figure 5.5 shows how such devices may be connected in a system. While typically only one device type is connected at a given substation, in some instances all three types of the devices may be deployed. For the applications of interest in this

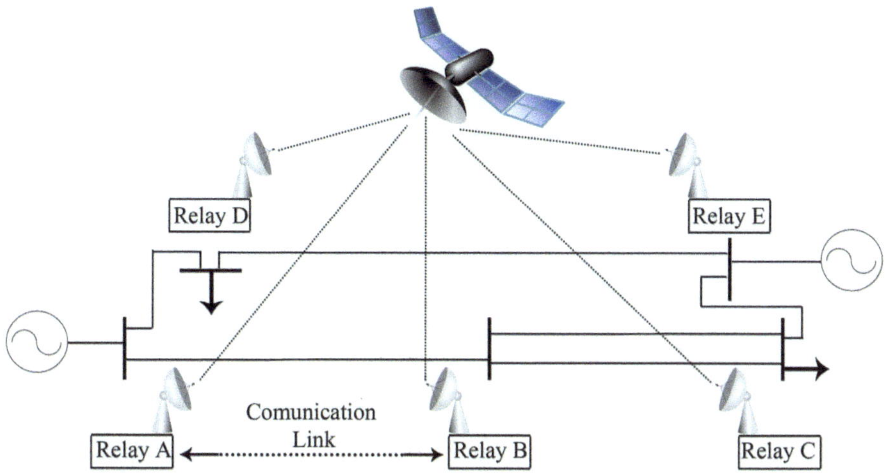

Fig. 5.4 Typical GPS synchronization connection of digital relays

discussion, at least one of such devices needs to be available at the points of interest and with the corresponding capability to provide either synchronized samples or synchronized phasors.

Further details of such multiple device connections in a given substation are given in Fig. 5.6. As may be noted, a variety of communication switches and GPS time signal sources available at different substations may be interconnected and made available for remote access from various remote locations.

5.3 The Use of Synchronized Samples from Two Ends

The use of synchronized samples for fault location entails an assumption of the model of the transmission line of interest: lumped (short line) or distributed parameters (long line) as shown in Fig. 5.7.

5.3.1 Theoretical Formulation

Details of this approach have been explored in many papers and several Ph.D. dissertations [5–12]. The transmission line considered in this section has a length l and is homogeneous to simplify the presentation. The assumptions can easily be extended to include distributed parameter transmission line models with capacitance included. For the simplified consideration, conductance and capacitance are neglected. A one-line representation of the three-phase line, with a fault at distance z from end S, is represented in Fig. 5.8.

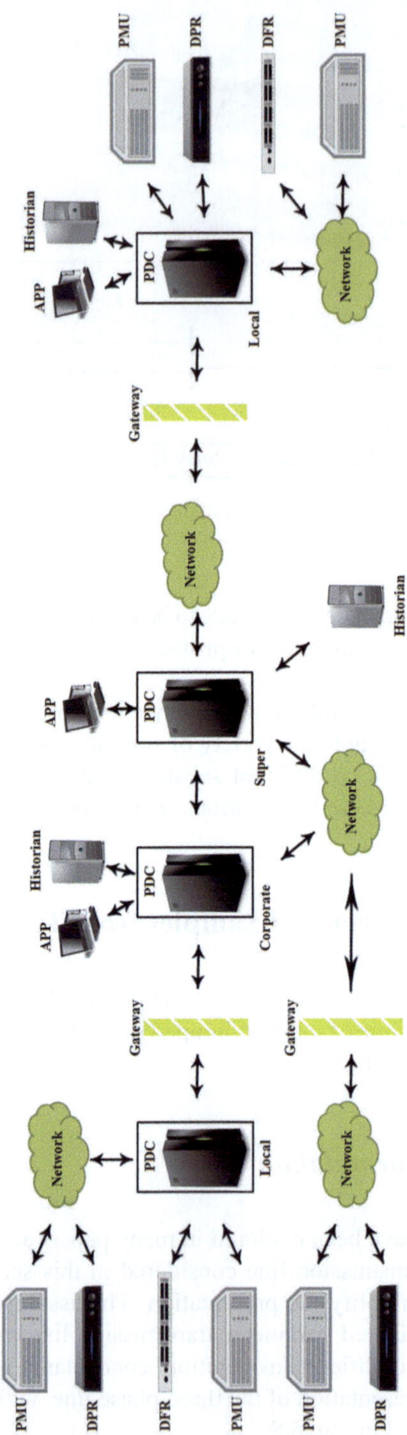

Fig. 5.5 Remote access to multiple devices providing synchronized samples and/or phasors. *PMU* phasor measurement unit, *DPR* digital protective relay, *DFR* digital fault recorder

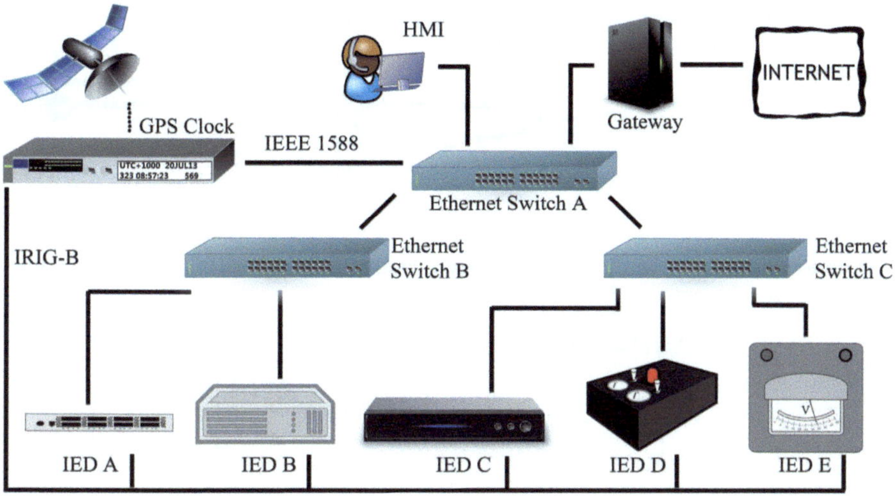

Fig. 5.6 IEDs with equipment providing synchronized samples and/or phasors. *IED* intelligent electronic device

Fig. 5.7 Transmission line model options

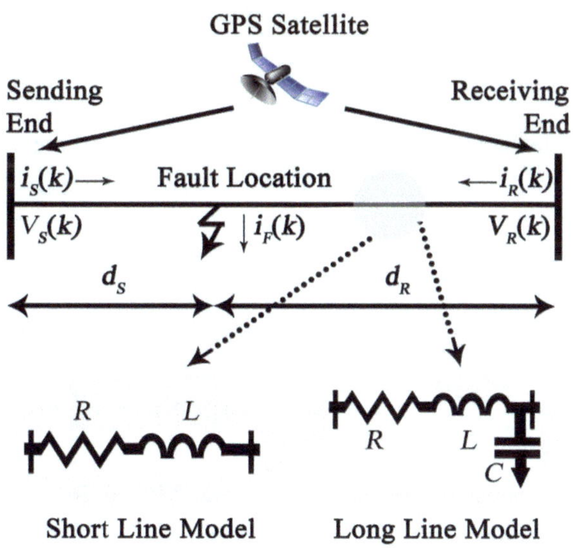

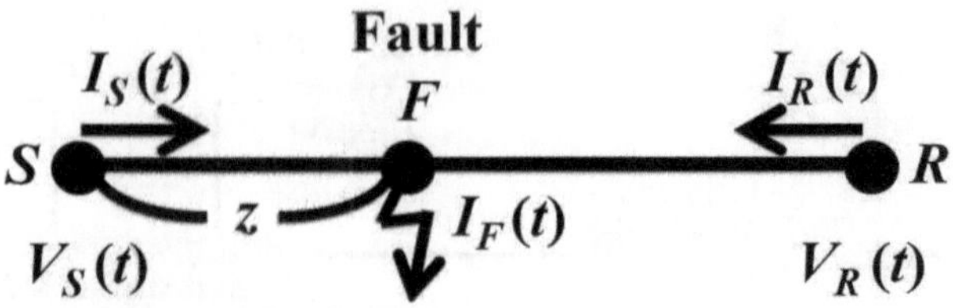

Fig. 5.8 One line representation of the three-phase transmission line

The location of the fault is based on two vectors defined as:

$$\begin{aligned}\Delta I(t) &= I_S(t) + I_R(t)\\ \Delta V(t) &= V_S(t) - V_R(t) + l\left[RI_R(t) + L\frac{dI_R(t)}{dt}\right],\end{aligned} \tag{5.1}$$

where $V_S(t)$ and $V_R(t)$ are vectors of phase voltages; $I_S(t)$, $I_R(t)$, and $I_F(t)$ are vectors of phase currents; S and R designate sending and receiving end of a transmission line, respectively; and R and L are matrices of self and mutual line parameters.

Under normal operating conditions, the fault current $I_F(t)$ is zero. As a consequence of Kirchhoff's current and voltage laws, the above vectors are equal to zero as:

$$\begin{aligned}\Delta I(t) &= 0\\ \Delta V(t) &= 0^{\cdot}\end{aligned} \tag{5.2}$$

If the line is faulted, the values of these vectors are:

$$\begin{aligned}\Delta I(t) &= I_F(t)\\ \Delta V(t) &= x\left[RI_F(t) + L\frac{dI_F(t)}{dt}\right].\end{aligned} \tag{5.3}$$

The fault current may be eliminated from Eq. (5.3) leading to:

$$\Delta V(t) - x\left[R\Delta I(t) + L\frac{d}{dt}\Delta I(t)\right] = 0. \tag{5.4}$$

The values of the vectors $\Delta V(t)$ and $\Delta I(t)$ in Eq. (5.1) at times n/f_s, where f_s, is the sampling frequency, can be calculated from current and voltage samples:

$$\begin{aligned}\Delta I_n &= I_{Sn} + I_{Rn}\\ \Delta V_n &= V_{Sn} - V_{Rn} + l\left[RI_{Rn} + f_s L(I_{Rn} - I_{Rn-1})\right].\end{aligned} \tag{5.5}$$

Here V_{Sn}, V_{Rn}, I_{Sn}, and I_{Rn} denote vectors of samples taken synchronously at moments n/f_s. It should be noted that the expression for ΔV_n is an approximate one, since the current derivative cannot be measured. Here, the derivative is approximated with "backward" approximation.

The discrete version of Eq. (5.4) is obtained using the same approximation for the current derivative as:

$$\Delta V_n - x\varphi_n = 0, \tag{5.6}$$

where φ_n can be expressed as

$$\varphi_n = R\Delta I_n + f_s L(\Delta I_n - \Delta I_{n-1}). \tag{5.7}$$

The fault location is based on Eqs. (5.6) and (5.7). In these equations, ΔV_n, ΔI_n and line parameters are known, while the distance x to the fault has to be determined. The number of available scalar equations is equal to $3N$, if the samples are taken at $N+1$ instant. Since the system of equations is an over-determined one, the most suitable way is to apply the minimum least square method. This method provides the following best estimate of x:

$$x = \frac{\sum_{n=2}^{N+1} \langle \Delta V_n, \varphi_n \rangle}{\sum_{n=2}^{N+1} \langle \varphi_n, \varphi_n \rangle}, \tag{5.8}$$

where $\langle , \rangle$ denotes the scalar product of vectors.

All the principles of fault analysis presented here are based on fundamental electrical laws only. Derived expressions are valid for any set of voltages and currents at transmission line ends. Consequently, voltages do not need be balanced; some phase voltages may even be zero. This makes the method completely independent of power system operating conditions. The transmission line does not have to be transposed. No assumptions about fault currents need to be imposed. The fault resistance may arbitrarily vary in time. Even an inductive component in the fault impedance may be present. This ability to accommodate different operating modes and models of the power system is a unique feature of the method presented.

Since in the selected example the conductance and capacitance are neglected, the method, as derived, is most suitable for shorter lines. As mentioned earlier, the techniques can be easily extended to include the long-line representation as well. The inevitable approximation of the current derivative is also an error source [13]. Further details are focused on determining the fault type, which is essential for applying the fault location equations, and on distributed parameter transmission line application, which requires particular adjustments in the algorithm use.

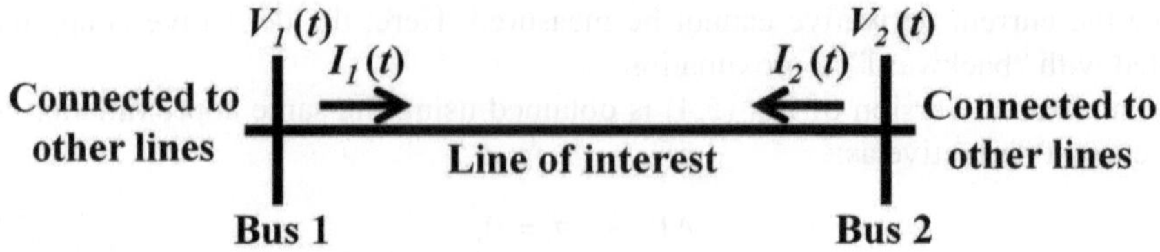

Fig. 5.9 Transmission line with measurements at two ends

5.3.2 Implementation Details and Results

The implementation consists of the following steps:

- Fault detection and classification [5]
- Fault location calculation [11, 12]

5.3.2.1 Fault Detection and Classification

In Fig. 5.9, $V_1(t)$ and $I_1(t)$ represent voltage and current measured at one end (bus 1) of the line at time instant *t*. Similarly, $V_2(t)$ and $I_2(t)$ represent voltage and current measured at the other end (bus 2) of the line at same time instant *t*. Currents are measured in the assumed direction, as shown in Fig. 5.9. All voltage and currents are single phase quantities. Voltage and current at bus 1 are shown in Eq. (5.9):

$$\begin{aligned} V_1(t) &= V_{1m} \cos \omega t \\ I_1(t) &= I_{1m} \cos(\omega t - \theta_1), \end{aligned} \tag{5.9}$$

where V_{1m} is the maximum value of voltage, I_{1m} is the maximum value of current, ω is the angular frequency, and θ_1 is the phase angle between $V_1(t)$ and $I_1(t)$.

The instantaneous power at bus 1 is shown in Eq. (5.10):

$$\begin{aligned} & P_1(t) \\ &= V_1(t) \times I_1(t) \\ &= V_{1m} I_{1m} \cos \omega t \cos\left(\omega t - \theta_1\right) \\ &= \frac{V_{1m} I_{1m}}{2}\left[\cos(2\omega t - \theta_1) + \cos \theta_1\right] \\ &= \frac{V_{1m} I_{1m}}{2}(\cos 2\omega t \cos \theta_1 + \sin 2\omega t \sin \theta_1 + \cos \theta_1) \\ &= P_{1m}(\cos 2\omega t + 1)\cos \theta_1 + P_{1m} \sin 2\omega t \sin \theta_1. \end{aligned} \tag{5.10}$$

The voltage and current at bus 2 are shown in Eq. (5.11):

$$
\begin{aligned}
V_2(t) &= V_{2m} cos(\omega t - \delta) \\
I_2(t) &= I_{2m} cos(\omega t - \delta - \theta_2)
\end{aligned},
\tag{5.11}
$$

where V_{2m} is the maximum value of voltage, I_{2m} is the maximum value of current, θ_2 is the phase angle between $V_2(t)$ and $I_2(t)$, and δ is the phase angle difference between buses 1 and 2.

The instantaneous power at bus 2 is shown in Eq. (5.12):

$$
\begin{aligned}
& P_2(t) \\
&= V_2(t) \times I_2(t) \\
&= V_{2m} I_{2m} \cos(\omega t - \delta)\cos(\omega t - \delta - \theta_2) \\
&= P_{2m}\left[\cos 2(\omega t - \delta) + 1\right]\cos\theta_2 + P_{2m}\sin 2(\omega t - \delta)\sin\theta_2
\end{aligned}.
\tag{5.12}
$$

Now, with the assumed direction of currents, the magnitude of $I_2(t)$ is negative before fault and positive after fault (change of direction).

For the unfaulted situation, the instantaneous powers are shown in Eq. (5.13):

$$
\begin{aligned}
P_1^u(t) &= P_{1m}^u(\cos 2\omega t + 1)\cos\theta_1^u + P_{1m}^u \sin 2\omega t \sin\theta_1^u \\
P_2^u(t) &= -P_{2m}^u\left[\cos 2(\omega t - \delta) + 1\right]\cos\theta_2^u - P_{2m}^u \sin 2(\omega t - \delta)\sin\theta_2^u
\end{aligned}.
\tag{5.13}
$$

After the fault, the instantaneous powers are shown in Eq. (5.14):

$$
\begin{aligned}
P_1^f(t) &= P_{1m}^f(\cos 2\omega t + 1)\cos\theta_1^f + P_{1m}^f \sin 2\omega t \sin\theta_1^f \\
P_2^f(t) &= P_{2m}^f\left[\cos 2(\omega t - \delta) + 1\right]\cos\theta_2^f + P_{2m}^f \sin 2(\omega t - \delta)\sin\theta_2^f
\end{aligned}.
\tag{5.14}
$$

If both before the fault and after the fault, power factor angles are lagging, i.e., $\theta_1^u > 0$, $\theta_2^u > 0$, $\theta_1^f > 0$, and $\theta_2^f > 0$, then the inequalities in Eqs. (5.15) and (5.16) should be satisfied. The inequalities in Eqs. (5.15) and (5.16) show the conditions before and after the fault, respectively:

$$
\begin{aligned}
\left|P_1^u(t)\right| &> 0 \\
\left|P_2^u(t)\right| &< 0
\end{aligned},
\tag{5.15}
$$

$$
\begin{aligned}
\left|P_1^f(t)\right| &> 0 \\
\left|P_2^f(t)\right| &> 0
\end{aligned}.
\tag{5.16}
$$

If before the fault power factor angles are leading, and after the fault power factor angles are lagging, i.e., $\theta_1^u < 0, \theta_2^u < 0, \theta_1^f > 0$, and $\theta_2^f > 0$ then the inequalities in

Eqs. (5.17) and (5.18) should be satisfied. The inequalities in Eqs. (5.17) and (5.18) show the conditions before and after the fault, respectively,

$$\begin{gathered}\left|P_1^u(t)\right|>0 \text{ if } (\cos 2\omega t+1)\cos\theta_1^u>\sin 2\omega t \sin\theta_1^u \\ \left|P_2^u(t)\right|<0 \text{ if } [\cos 2(\omega t-\delta)+1]\cos\theta_2^u>\sin 2(\omega t-\delta)\sin\theta_2^u\end{gathered}, \tag{5.17}$$

$$\begin{gathered}\left|P_1^f(t)\right|>0 \\ \left|P_2^f(t)\right|>0\end{gathered}. \tag{5.18}$$

This can be shown for all combinations of lagging and leading power factor angles before and after fault, $\left|P_1^u(t)\right|>0, \left|P_2^u(t)\right|<0$ and $\left|P_1^f(t)\right|>0, \left|P_2^f(t)\right|>0$, if one or some of the inequalities in Eq. (5.19) are true:

$$\begin{gathered}(\cos 2\omega t+1)\cos\theta_1^u>\sin 2\omega t\sin\theta_1^u \\ [\cos 2(\omega t-\delta)+1]\cos\theta_2^u>\sin 2(\omega t-\delta)\sin\theta_2^u \\ (\cos 2\omega t+1)\cos\theta_1^f>\sin 2\omega t\sin\theta_1^f \\ [\cos 2(\omega t-\delta)+1]\cos\theta_2^f>\sin 2(\omega t-\delta)\sin\theta_2^f\end{gathered}. \tag{5.19}$$

Under small values of power factor angles, all of the inequalities in (5.19) are satisfied. In general, in transmission systems, before fault power factor angles are very small and after fault they are lagging, which is sufficient to conclude that inequalities in Eqs. (5.15) and (5.16) are true.

Therefore, this unique feature of instantaneous power under different types of faults helps detect and classify faults without using any threshold. This feature is observed only on the faulted phases.

The following sections illustrate how this concept may be used to detect and classify single phase-to-ground faults and phase-to-phase faults. Further examples showing how other types of faults may be detected may be found in [5] and [11].

Single Phase-to-Ground Fault In the case of a single phase-to-ground fault ("*ag*"), the plot $P_1(t)$ of and $P_2(t)$ with respect to time is shown in Fig. 5.10a–c. Right after the fault inception (0.02 s), $\left|P_{1a}^f(t)\right|$ and $\left|P_{2a}^f(t)\right|$ are greater than zero in phase "*a*" while for the other two phases $\left|P_{1b}^f(t)\right|>0, \left|P_{2b}^f(t)\right|<0$ and $\left|P_{1c}^f(t)\right|>0, \left|P_{2c}^f(t)\right|<0$.

To represent this feature mathematically, the signum function is used, which is defined as in Eq. (5.20):

$$\operatorname{sgn}(x)=\begin{cases}-1, x<0 \\ 0, x=0 \\ 1, x>0\end{cases}. \tag{5.20}$$

$\operatorname{sgn}(p_1(t))$ and $\operatorname{sgn}(p_2(t))$ are calculated for each phase and the difference $P\operatorname{sgn}(t)=\operatorname{sgn}(P_1(t))-\operatorname{sgn}(P_2(t))$ for each phase is plotted in Fig. 5.10d–f.

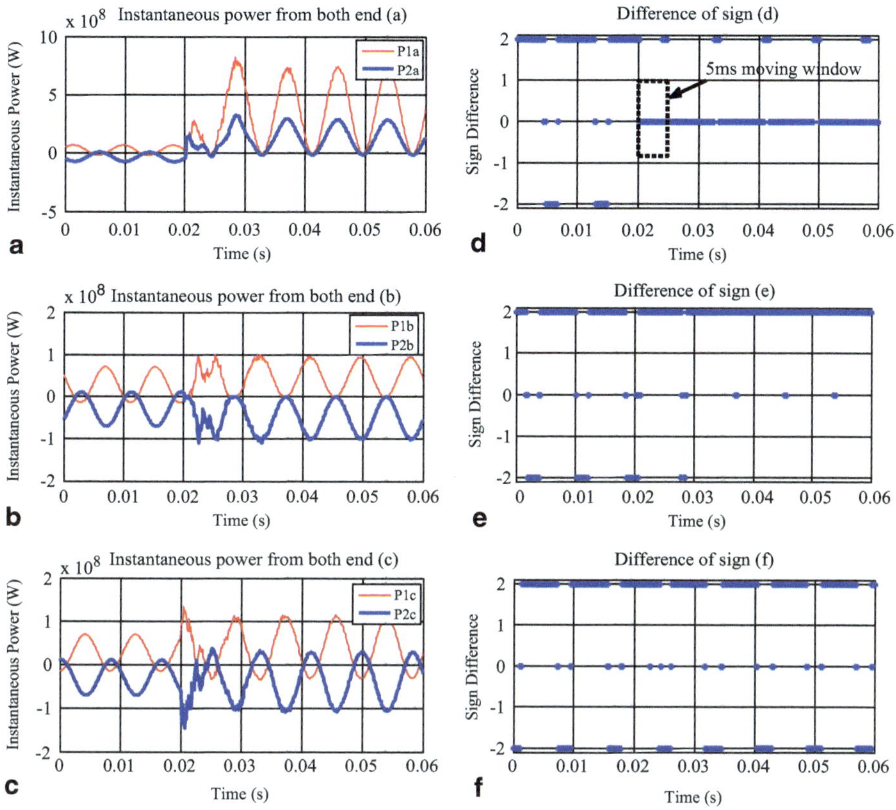

Fig. 5.10 (a–c): $P_1(t)$ and $P_2(t)$ with respect to time for "*ag*" fault; (d–f): $P\text{sgn}(t)$ with respect to time for "*ag*" fault

Theoretically, before the fault this difference $P\text{sgn}(t)$ should be ± 2 and after fault $P\text{sgn}(t)$ should be zero, but due to transients and noise present in the measurements, some outliers are also present. It is clear from Fig. 5.10d–f that on phase "*a*" $P\text{sgn}(t)$ becomes almost zero after fault while the other phases remain unchanged. The time instant at which this change occurs is the fault instant. A moving window (5 ms taken in this example) is used to check whether at least 90 % (due to presence of outliers) of $P\text{sgn}(t)$ are zero, which indicates the instant faulted phase (phase "*a*" here) experienced a fault.

Phase-to-Phase Fault In the case of a phase-to-phase fault ("*ab*"), the plot of $P_1(t)$ and $P_2(t)$ with respect to time is shown in Fig. 5.11a–c. Right after the fault inception (0.02 s), $|P^f_{1a}(t)|$ and $|P^f_{2a}(t)|$ are greater than zero in phase "*a*," and $|P^f_{1b}(t)|$ and $|P^f_{2b}(t)|$ are greater than zero in phase "*b*" while $|P^f_{1c}(t)| > 0$ and $|P^f_{1c}(t)| < 0$ in phase "*c*." The difference $P\text{sgn}(t) = \text{sgn}(P_1(t)) - \text{sgn}(P_2(t))$ for each phase is plotted in Fig. 5.11d–f. It is clear that after fault, $P\text{sgn}(t)$ is almost zero for both phases "*a*" and "*b*." Note the same logic used for single phase-to-ground faults may be used to detect phase-to-phase faults. More test results are available in [5].

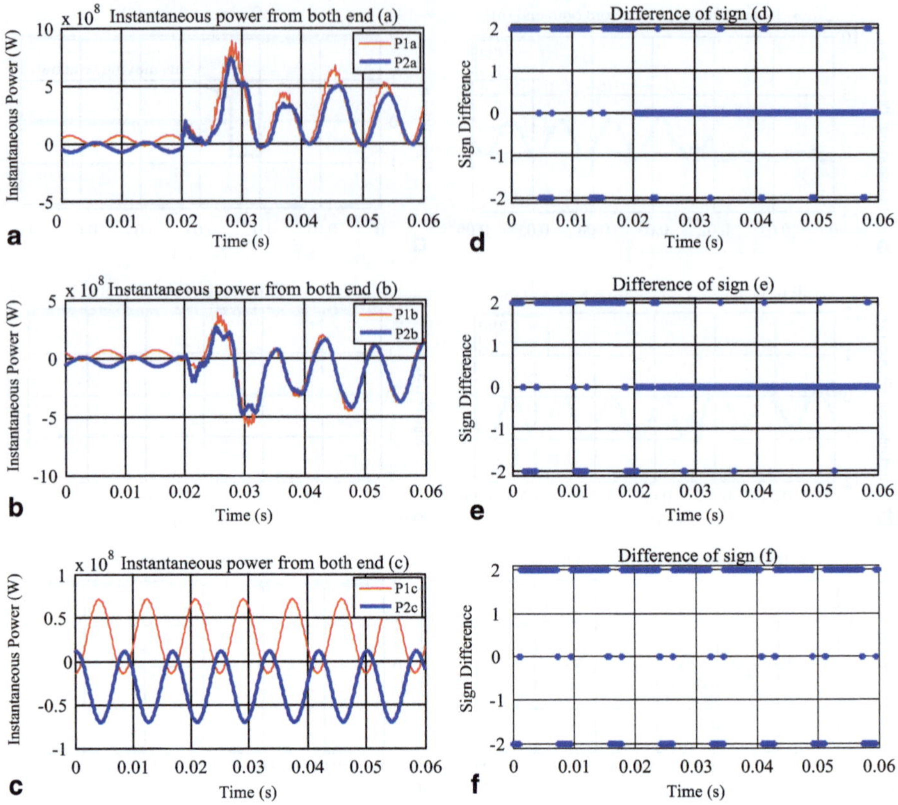

Fig. 5.11 (a–c) $P_1(t)$ and $P_2(t)$ with respect to time for "*ab*" fault; (d–f) $P\mathrm{sgn}(t)$ with respect to time for "*ab*" fault

5.3.2.2 Fault Location Using Distributed Parameter Model

Once the fault classification scheme detects a fault in a transmission line, the synchronized sampling-based fault location scheme, originally proposed in [8] and then further explored in [11] and [12], is used to locate exact location of the fault using fault detection and classification outputs. The proposed fault location method requires high sampling rate of data, and yet the method used for fault detection and classification requires lower sampling rate. A spline interpolation technique [13] is used to introduce samples in between two adjacent samples of original voltage and current waveforms, thereby increasing the sampling rate for input data.

A segment Δx of transmission line is shown in Fig. 5.12. Transmission line partial differential equations are shown in Eq. (5.21):

$$\begin{aligned} \frac{\partial v(x,t)}{\partial x} &= -Ri(x,t) - L\frac{\partial i(x,t)}{\partial t} \\ \frac{\partial i(x,t)}{\partial x} &= -Gv(x,t) - C\frac{\partial v(x,t)}{\partial t}, \end{aligned} \tag{5.21}$$

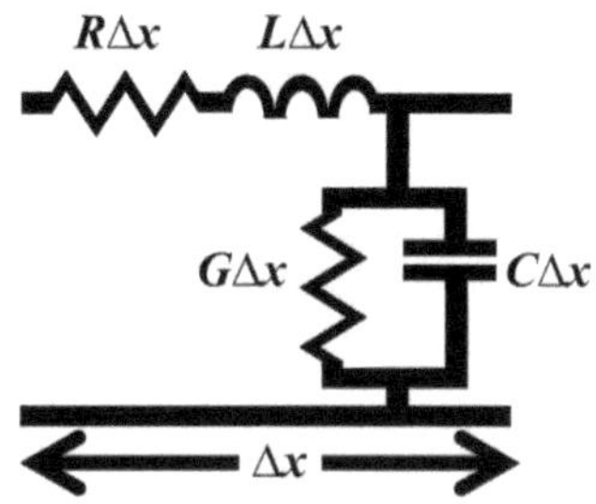

Fig. 5.12 Transmission line segment of length

where $v(x, t)$ and $i(x, t)$ are the instantaneous voltage and current at any point X along the line, at distance x from one end of the line.

These hyperbolic partial differential equations are solved using the method of characteristics [14]. The transmission line equations are transformed into an equivalent set of ordinary differential equations. These ordinary differential equations apply only in certain directions in the (x, t) plane, known as characteristic directions.

Define characteristics impedance $Z_C = \sqrt{\frac{L}{C}}$ and propagation constant $\gamma = \sqrt{LC}$ and substituting into Eq. (5.21), the governing relations shown in Eq. (5.22) are obtained:

$$\begin{aligned}\left(\frac{\partial}{\partial x} + \gamma\frac{\partial}{\partial t}\right)(v + Z_C i) &= -Ri - GZ_C v \\ \left(\frac{\partial}{\partial x} - \gamma\frac{\partial}{\partial t}\right)(v - Z_C i) &= -Ri + GZ_C v\end{aligned}. \tag{5.22}$$

Considering the function $f(x, t)$ as shown in Eq. (5.23), $\left(\frac{\partial}{\partial x} + a\frac{\partial}{\partial t}\right)$ can be treated as a differential operator which gives total derivative of $f(x, t)$ with respect to x in the direction of $\frac{dt}{dx} = a$ in the (x, t) plane:

$$\frac{df}{dx} = \frac{\partial f}{\partial x} + \frac{dt}{dx}\frac{\partial f}{\partial t} = \left(\frac{\partial}{\partial x} + \frac{dt}{dx}\frac{\partial}{\partial t}\right)f. \tag{5.23}$$

Therefore, the set of ordinary differential equations are obtained as shown in Eq. (5.24):

$$\begin{aligned}\frac{d}{dx}(v + Z_C i) &= \frac{dv}{dx} + Z_C\frac{di}{dx} = -Ri - GZ_C v \text{ along } \frac{dt}{dx} = +\gamma \\ \frac{d}{dx}(v - Z_C i) &= \frac{dv}{dx} - Z_C\frac{di}{dx} = -Ri + GZ_C v \text{ along } \frac{dt}{dx} = -\gamma\end{aligned}. \tag{5.24}$$

Figure 5.13 shows the ordinary differential Eq. (5.23) in the (x, t) plane. Define distance segment as $\Delta x = x_j - x_{j-1}$ when Δx is the distance that the wave travels with

Fig. 5.13 Set of ordinary differential equations in (*x*, *t*) plane

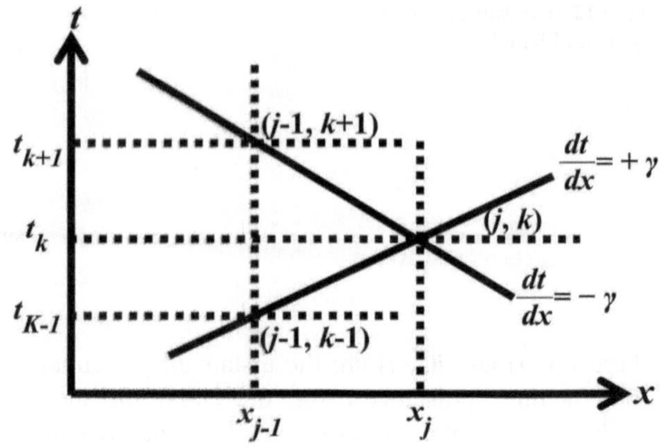

a sampling time step, $\Delta t = t_{k+1} - t_k = t_k - t_{k-1}$, where k is the present sample point: $\Delta t = \gamma \Delta x$.

Points p, s, and r are defined in the (x, t) plane: $p = (j, k)$, $s = (j-1, k-1)$, $r = (j-1, k+1)$. From Eq. (5.24), one obtains

$$\begin{aligned}
&\left(\frac{v(p)-v(s)}{\Delta x}\right)+Z_C\left(\frac{i(p)-i(s)}{\Delta x}\right)=-R\left(\frac{i(p)+i(s)}{2}\right)-GZ_C\left(\frac{v(p)+v(s)}{2}\right)\\
&\left(\frac{v(r)-v(p)}{\Delta x}\right)-Z_C\left(\frac{i(r)-i(p)}{\Delta x}\right)=-R\left(\frac{i(r)+i(p)}{2}\right)+GZ_C\left(\frac{v(r)+v(p)}{2}\right)
\end{aligned}. \tag{5.25}$$

Adding and subtracting the two equations in Eq. (5.25), the following relationships are obtained:

$$\begin{aligned}
&\left(1+\frac{R\Delta x}{2Z_C}\right)i(p)=\left(\frac{1}{2}-\frac{R\Delta x}{4Z_C}\right)\left(i(r)+i(s)\right)+\left(\frac{G\Delta x}{4}-\frac{1}{2Z_C}\right)\left(v(r)-v(s)\right)\\
&\left(1+\frac{GZ_C\Delta x}{2}\right)v(p)=\left(\frac{R\Delta x}{4}-\frac{Z_C}{2}\right)\left(i(r)-i(s)\right)+\left(\frac{1}{2}-\frac{GZ_C\Delta x}{4}\right)\left(v(r)+v(s)\right)
\end{aligned}. \tag{5.26}$$

Assuming $G=0$, Eq. (5.26) can be rewritten as:

$$\begin{aligned}
&\left(1+\frac{R\Delta x}{2Z_C}\right)i(j,k)=\left(\frac{1}{2}-\frac{R\Delta x}{4Z_C}\right)\left(i(j-1,k+1)+i(j-1,k-1)\right)+\frac{1}{2Z_C}\left(v(j-1,k+1)-v(j-1,k-1)\right)\\
&v(j,k)=\left(\frac{R\Delta x}{4}-\frac{Z_C}{2}\right)\left(i(j-1,k+1)-i(j-1,k-1)\right)+\frac{1}{2}\left(v(j-1,k+1)+v(j-1,k-1)\right).
\end{aligned} \tag{5.27}$$

Since the explicit form of the fault location cannot be obtained, an indirect approach is used to calculate the final fault location as described in the following steps:

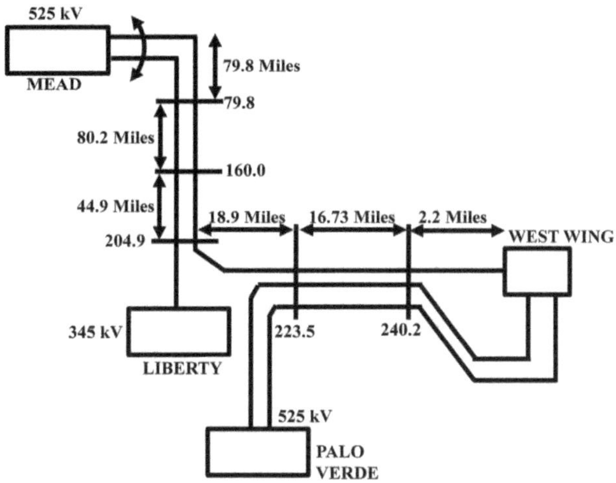

Fig. 5.14 One-line diagram of the WAPA test system

1. Discretize the line into equal segments with length of Δx and build a voltage profile for each point calculating from sending end and receiving end, respectively.
2. Locate the approximate fault point by finding the point that has the minimum square of voltage difference calculated from both ends.
3. Build a short line model surrounding the approximate fault point, and refine the fault location using the algorithm based on lumped line parameters.

As the fault location method is based on the distributed parameter line model, it requires very high sampling rates for voltage and current signals. Since the fault detection and classification method does not require high sampling rates, the spline interpolation technique is used between input samples to achieve higher sampling rate appearance for fault location application.

5.3.2.3 Test Study

The following discussion is related to a test study using a few real-life test scenarios reported in [6] and [12]. The testing of the algorithm was then carried out using data generated from the electromagnetic transients program (EMTP) simulation of a real power system belonging to Western Area Power Administration (WAPA). The one-line diagram of the system is shown in Fig. 5.14.

The line of interest is the 525 kV line from the Mead to Westwing substation. The line is 242.4 miles long and is mutually coupled to the 345 kV Mead–Liberty line for 204.9 miles, starting from Mead. There is also a 16.73 mile section, that is mutually coupled to two 525 kV lines starting at Palo Verde and terminating at Westwing. The Mead–Westwing and Mead–Liberty lines are equipped with series compensation capacitors at each end. The capacitors are protected by metal oxide varistors (MOVs). For a detailed description of the system, please refer to [15]. The voltage and current data are acquired from the Mead 525 and 345 kV buses, the

Table 5.1 Maximum fault location error (%) *a–g* fault

Actual location	Error and computed location			
(miles from Mead)	PC	PN	SC	SN
40.0	0.355	0.708	1.498	1.604
	39.15	38.28	43.63	43.881
79.8	0.367	0.437	0.246	1.604
	80.68	80.86	80.40	83.21
160.0	1.076	1.128	1.084	1.196
	162.61	162.73	157.37	162.9
223.5	2.727	2.041	1.570	1.339
	230.11	228.45	227.30	226.75

Table 5.2 Maximum fault location error (%) *b–c* fault

Actual location	Error and computed location			
(miles from Mead)	PC	PN	SC	SN
40.0	1.650	1.833	0.402	0.098
	36.00	35.56	39.03	39.76
79.8	3.791	3.283	3.513	0.913
	70.61	71.84	71.28	77.59
160.0	3.013	1.526	2.880	0.848
	152.70	156.30	153.02	157.94
223.5	1.856	0.757	0.828	1.058
	228.00	225.34	221.49	220.93

Westwing 525 kV bus, and the Liberty 345 kV bus, to give a total of 4 three-phase voltages and 4 three-phase currents. No data are available from the Palo Verde–Westwing double circuit line.

On this system, extensive EMTP simulations were carried out, with the following variations in the simulation conditions [16]:

- Fault Locations: Faults were introduced at 40.0, 79.8, 160.0, and 223.5 miles from Mead.
- Fault Types: Four types of faults were considered; phase "*a*" to ground, phase "*b*" to phase "*c*," phase "*b*" to phase "*c*" to ground and three phase-to-ground.
- Fault Incidence Angle: 0° and 90°.
- Fault Impedance: 3 and 50 ohms.
- Series Capacitors: Capacitors In and Capacitors Bypassed.
- Instrument Transformers: Data collected from the primary and from the secondary of the instrument transformers. The instrument transformer models used in this study were developed earlier [17, 18].

In total, there are four fault locations, four fault types, two fault incidence, two fault impedance, two capacitors, and two instrument transformers. This gives a total of 256 test cases.

Tables 5.1 and 5.2 show the results for phase "*a*" to ground faults and phase "*b*" to "*c*" faults, respectively. Each column in the tables corresponds to the following four system conditions respectively:

- PC: Data from Primary, with Series Capacitors
- PN: Data from Primary, without Series Capacitors
- SC: Data from Secondary, with Series Capacitors
- SN: Data from Secondary, without Series Capacitors

The error % is shown in each cell of the table, and below it is the distance as computed by the fault location algorithm. The error % is the worst-case error of four different fault scenarios which are obtained for the two fault impedances and two incidence angles. The error is calculated by the formula given below. The sampling frequency is 20 kHz. All distances are measured in miles from Mead substation. The single line to ground fault shows a better accuracy than the multiphase fault for most of the testing scenarios. The accuracy of the fault location when the data are measured at the primary of the instrument transformers (scenarios PC and PN) is a little better than when the data are measured at the secondary (scenarios SC and SN), for a number of cases.

The largest error occurs for the phase "*a*" to ground fault at 223.5 miles from Mead, with the series capacitors included in the simulation and with the data measured at the primary. Other than this, there is no clear pattern in the error percentages. The testing also showed that the maximum errors were not associated with one particular fault incidence angle or fault impedance. Errors for the phase "*b*" to "*c*" fault are higher in general, than the errors for the single line to ground fault. When data are measured from the secondary of the instrument transformers, the error percentages are lower than the cases when data are measured from the primary. The largest error is 3.791 % for a fault at 79.8 miles from Mead, with the series capacitors included and data taken from the primary. This corresponds to an error in the fault location of 9.19 miles. The lowest errors are seen for faults at 40.0 miles from Mead. At 79.8 and 160.0 miles from Mead, the errors increase, and then drop again at 223.5 miles. As in the case of the phase "*a*" to ground fault, there was no association between the maximum error and any particular incidence angle or fault impedance. There are several factors affecting the fault location accuracy. One of the main factors affecting the fault location accuracy is the sampling frequency of the data. The sampling interval determines the length of each discrete segment of the long transmission line. In our algorithm, however, the fault location does not stop with determining the approximate point. It is further refined by applying the short line algorithm on a segment of the line around the approximate fault point. Therefore, the sampling frequency does not play as important a role as it does when traveling wave methods are used [19]. However, the sampling frequency must be high enough to ensure that the refinement of the fault location is carried out on a segment that is reasonably short. In our case, this segment is around 16 miles, at a sampling frequency of 20 kHz.

The main factors affecting the accuracy of the fault location are:

- **Transposition points in the power system:** In Fig. 5.14, transposition points are seen at 79.8, 160.0, 204.9, and 223.5 miles in the line. These points are sources of reflected traveling waves, which affect the terminal voltage and current data. The proposed algorithm however, considers the line to be made up of three homogeneous sections: Mead to Fault Point, Fault Point to end of Mutual Coupling (204.9 miles), and from 204.9 miles to Westwing. If the fault is outside the mutually coupled section, then the algorithm assumes that the line is made up of two homogeneous sections. The reflections from the transposition points affect the accuracy of the voltage and current reconstruction at the points adjacent to the approximate point, which in turn affects the accuracy of the fault location.

- **Mutual coupling with the Palo Verde lines:** There is one section 16.73 miles long, where the Mead–Westwing line is coupled with two 525 kV line from Palo Verde. This coupling is not considered by the algorithm and is another source of error.
- **Impact of series capacitors or surge arresters**: Modeling components like series capacitors and surge arresters in the fault location algorithm did not affect the accuracy of the algorithm. As can be seen from the results, there is no clear trend to indicate if the presence or absence of the series capacitors in the simulations, affected the errors in any way, since all voltage and current measurements were taken from the line side of the circuit breakers.

5.3.3 Features and Benefits

The technique presented in this section has the following features:

- The technique proposed is based on an off-the-shelf recording IEDs incorporating GPS receiver providing synchronized sampling at two ends of a transmission line, which significantly improves quality of data.
- Due to availability of synchronously sampled data, a unique time domain approach for fault analysis is feasible, which is not sensitive to fault resistance or changes in the fault arc characteristics.
- The new fault analysis approach is extremely fast, selective, and accurate providing, at the same time, low sensitivity to sources of errors coming from a variety of power system conditions.
- Since future developments are leaning towards entirely automated fault detection, classification, and location schemes, the technique proposed here is amenable to an easy implementation of such fault analysis in a compact form.

5.4 The Use of Sparse Synchronized Phasor Measurement for Fault Location

The use of synchronized phasors measured at two ends of a transmission line has been explored in the literature [20, 21]. However, installing recording devices (DFRs as an example) at the ends of all the transmission lines may not be feasible due to cost, or due to lack of adequate measurements as is the case with tapped lines

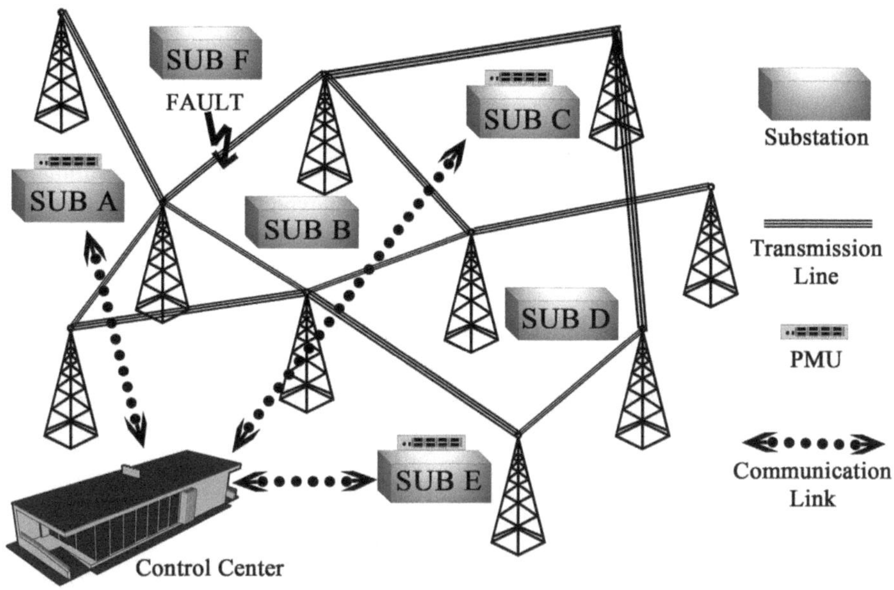

Fig. 5.15 The use of sparse phasor measurements for fault location. *PMU* phasor measurement unit

due to unavailability of instrument transformers. Although protective relays exist on every transmission line and isolate the faulted line by opening associated circuit breakers when sensing fault immediately, some of them may still be electromechanical, which do not have capability to record measurements. As a result, in some cases it may happen that there are no recordings available at line ends close to a fault. System-wide sparse measurement-based fault location method can be applied in such instances as shown in Fig. 5.15.

In the sparse measurement-based fault location method, synchronized phasor measurements from different substations located in the region where the fault has occurred are used. The measurements are considered sparse, as they may come from only some of many transmission line ends (substations) in the region. This method requires synchronization of the samples and extracted features (measurements), which may be obtained by using DFRs connected to GPS receivers. In addition to the sparse measurements, the technique also uses a short circuit program, which is initialized and tuned with SCADA PI Historian [22], power system model data, and measurements associated with the time of the fault occurrence.

5.4.1 Theoretical Formulation

The method uses waveform matching technique between the current and voltage phasors calculated from the samples of waveforms recorded in a substation (nearby the faulted line) and phasors simulated using short circuit calculation of possible fault locations [23]. Further details of the initial idea may be found in several references that explore different aspects of the proposed approach [11, 24–28]. The calculated and simulated phasors are compared while the location of the fault is changed in the short circuit program. This process of placing faults at different locations is repeated automatically until the difference between measured and simulated phasor values reaches a global optimum (minimum), which indicates that the fault location used in the short circuit program is the actual one in the field. The criteria for the minimal difference are based on a global optimization technique that uses genetic algorithm (GA).

In this approach, field-recorded samples of waveforms are used to calculate phasors, and they are in turn matched with the phasors obtained using short circuit study. The matching degree between the recorded and the simulated phasors can be formulated as:

$$f_c(x, R_F) = \sum_{K=1}^{N_V} r_{kV} \left| V_{ks} - V_{kr} \right| + \sum_{k=1}^{N_I} r_{kI} \left| I_{ks} - I_{kr} \right|, \tag{5.28}$$

where

$f_c\,(x, R_F)$ the cost function using phasors for matching,
x, R_F the fault location and fault resistance,
r_{kV}, r_{kI} weights for the errors of the voltages and currents, respectively,
V_{ks}, V_{kr} simulated and calculated from measurements during-fault voltages, respectively,
I_{ks}, I_{kr} simulated and calculated from measurements during-fault currents, respectively,
N_v, N_I total number of voltage and current phasors to be matched, respectively,
N_v, N_I the index of voltage or current phasors.

Ideally, when the simulated phasors and the phasor calculated from the recorded waveforms match completely, the cost function should become zero. In practical solutions, however, the cost function is not zero and therefore should be minimized using some mathematical optimization method. To obtain good phasor matching, the fault search range should be extensive. All possible faulted branches and fault resistance should be included in the search range, which makes the search two-dimensional and exhaustive. For a large system, multiple searches should be run in parallel and can be achieved using population-based optimization methods, such as GA [29]. The flowchart of this method is shown in Fig. 5.16.

5.4.2 Implementation Details and Results

Two issues are discussed in this section:

- Fault location calculation
- Fault resistance compensation

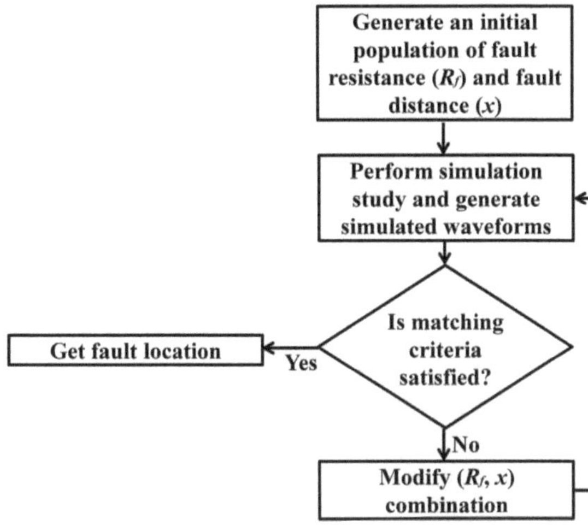

Fig. 5.16 Flowchart of sparse measurement algorithm

5.4.2.1 Fault Location Calculation

In the waveform matching approach described in Eq. (5.28), an initial fault location is first assumed and asserted in the model. Second, the simulation studies are configured according to the specified fault. Third, the simulation studies corresponding to the specified fault are carried out utilizing appropriate simulation tools. Fourth, simulated waveforms of quantities of interest are obtained. Fifth, the corresponding simulated waveforms are compared with recorded ones, and the matching degree of the simulated and recorded waveforms are evaluated by using appropriate criteria. Finally, the initial fault location is modified according to certain approaches, and then the process proceeds to the second step and continues. These steps are iterated until the simulated waveforms that best match the recorded ones are produced. The fault location will be determined as the one specified in the simulation studies for the case when generating simulated waveforms best match the recorded ones.

To evaluate the matching degree of the simulated and recorded waveforms, two different methods may be employed. The first one uses phasors for matching. The second one uses transients for matching. In this work, only the "phasor matching"-based approach for fault location is utilized.

The cost function $f_c(x, R_F)$ equals to zero theoretically when the phasors obtained from simulation studies exactly match those obtained from recorded waveforms. Therefore, for practical purposes, the best fault location estimate would be the one that minimizes the cost function. An appropriate optimization procedure needs to be selected to solve the minimized problem. Equation (5.28) can be converted into the maximum problem shown as follows:

$$f_f(x, R_F) = -f_c(x, R_F), \tag{5.29}$$

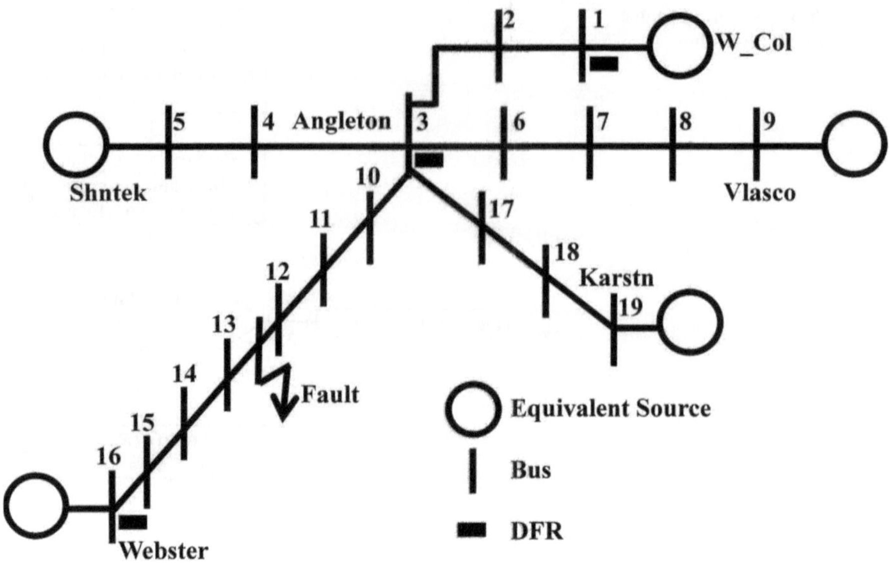

Fig. 5.17 Sample power system. *DFR* digital fault recorder

where $f_f(x, R_F)$ represents the fitness function when using phasor or magnitude for matching.

To investigate the nature of the fitness function, various simulation studies have been carried out to obtain the fitness value versus the fault location and resistance using the sample power system shown in Fig. 5.17.

The fitness value is obtained by specifying the faults with varying fault resistance on each line throughout the system, running simulations, and applying Eq. (5.29). When posing faults, the fault location changes in steps of 4 miles and the fault resistance in steps of 0.02 p.u. For each fault location and fault resistance, a corresponding fitness value can be obtained. To plot the fitness value versus the fault location and fault resistance, the value of the fault location needs to be defined.

First, the search sequence for all lines in the system should be defined. For example, suppose the faults are posed in the lines following the bus connection sequence: 1–2, 2–3, 3–4, 4–5, 3–6, 6–7, 7–8, 8–9, 3–10, 10–11, 11–12, 12–13, 13–14, 14–15, 15–16, 3–17, 17–18, 18–19 for the line sequence shown in Fig. 5.17. Bus 1 is defined as the initial bus for line 1, bus 2 is defined as the initial bus for line 2, and finally bus 18 is defined as the initial bus for the last line. The fault location is defined as the distance of the fault from the initial bus. When the fault location is determined to be at zero miles, the posed fault location is situated at bus 1. The next fault location is changed based on a preset step selected to be 4 miles. If the distance is smaller than the first line (1–2) length, the posed fault is located on the first line from the bus 1. Otherwise, the distance is compared with the sum of the first line (1–2) and the second line (2–3) length. If it is smaller, then the posed fault location is selected on the second line 4 miles away from bus 2.

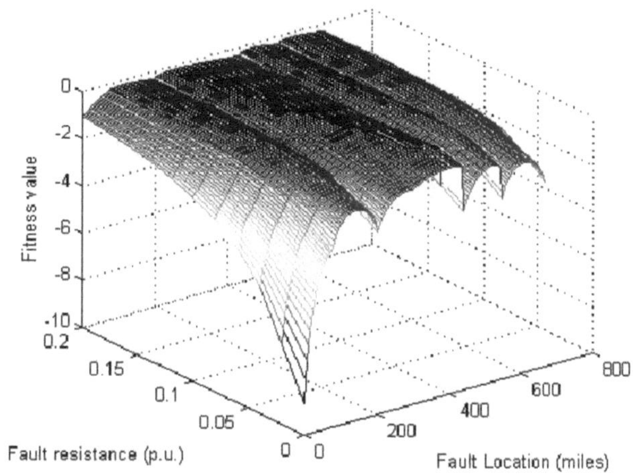

Fig. 5.18 The fitness surface for an *a–g* fault

Now assume that a phase "*a*" to ground fault with fault resistance 0.1 p.u. occurs on the line between buses 13 and 12, and is located 9.1 miles away from bus 13. The fault location is the distance from bus 1 to the fault point and it is obtained by adding those line lengths that contains lines 1–2, 2–3, 3–4, 4–5, 3–6, 6–7, 7–8, 8–9, 3–10, 10–11, 11–12, and the distance from bus 12 to the fault point. Therefore, the fault location would be at 312.7 miles according to the above definition. Details about the line lengths that lead to this calculation are given in [26]. If the recorded data are only available at bus 1, the fitness value versus the fault location and fault resistance for this specific fault can be obtained, as depicted in Fig. 5.18. In Fig. 5.18, the maximum fitness value occurs at point (312.7, 0.1), which is the optimal solution for phase A to ground fault. It is noted that the surface contains some local maximum and saddles. The surface is not regular. It contains saddle points and local maximum points. Simulation studies show similar characteristics for other types of faults. To find the global maximum point, optimization schemes in many shapes and sizes have been considered. The idea is fairly straightforward: within a finite search space or a discrete infinite search space, the search algorithm starts looking at objective function values at every point in the space, one at a time. The method is simple; however, it lacks efficiency.

Random search algorithms have achieved increased popularity because of the shortcoming of calculus-based and enumerative schemes. The GA is an example of a search procedure that uses random choice as a tool to guide a highly exploitative search through a coding of a parameter space. The research shows that the GA-based optimization approach is good at finding the globally optimal solution and avoiding the local optima; in other words, it is more robust than conventional search methods. Therefore, the GA is selected as the tool for finding the global optima in our study [29].

5.4.2.2 Fault Resistance Compensation

Since fault resistance is the most uncertain parameter, a correction scheme to compensate the effect of fault resistance is discussed next [30].

The correction scheme (shown in Fig. 5.19) compensates for the effect of fault resistance by using an optimization scheme over the sparse measurement method, which in turn selects the fault location and fault resistance pair that cause the output to be least sensitive to fault resistance variation.

The proposed corrected sparse measurement algorithm can be explained in the following step-by-step method:

Step 1 Initialization Generate a population of fault resistance values R_f (1 to N).

Step 2 Sparse Measurement Based Fault Location For each R_f, perform original sparse measurement algorithm and get fault location x (1 to N). Therefore, a pair of (x, R_f) is obtained for N number of cases.

Step 3 Estimation of Unknown Function g by Nonparametric Method There are N pairs of observations (x, R_f) for $i = 1{:}N$ that are used to estimate the unknown regression function $g(R_f)$ where fault location is expressed as a function of fault resistance, as shown in Eq. (5.30):

$$x = g(R_f) + \varepsilon, \tag{5.30}$$

where ε is assumed to be zero mean error. Therefore, the expected value of x given R_f is expressed in Eq. (5.31):

$$E\left(\frac{x}{R_f}\right) = g(R_f). \tag{5.31}$$

The true function g can be approximated by $\hat{g}$ using traditional nonparametric regression method [13].

Step 4 Calculate sensitivity to R_f The variance-based global sensitivity analysis method (ANOVA decomposition) is used [31]. The method is summarized below.

If a deterministic model $Z = f(y)$ is considered where y is a vector of input variables $y = (y_1, y_2, \ldots, y_k)$ and Z is the model output, $Z = f(y)$ can be decomposed into main effects and interactions, as shown in Eq. (5.32):

$$f(y) = f_0 + \sum_{i=1}^{k} f_i(y_i) + \sum_{i}\sum_{j>i} f_{ij}(y_i, y_j) + \ldots + f_{1,2,\ldots,k}(y_1, y_2, \ldots, y_k). \tag{5.32}$$

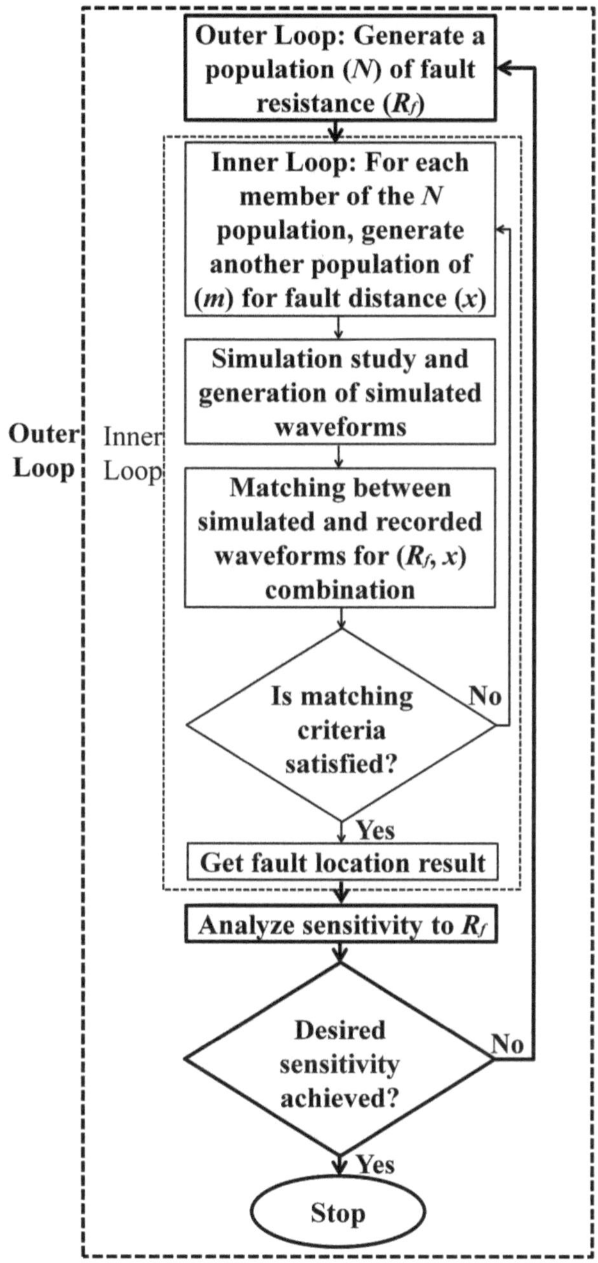

Fig. 5.19 Flowchart of corrected sparse measurement algorithm

If each term is chosen with zero mean as

$$\begin{aligned}&\int_0^1 f_i(y_i)dy_i = 0, \forall y_i, i = 1,2,\ldots,k\\&\int_0^1\int_0^1 f_{ij}(y_i,y_j)dy_i dy_j = 0, \forall y_i, y_j, i<j\ ,\\&\int_\Omega f_{1,2,\ldots,k}(y_1,y_2,\ldots,y_k)dy_1 dy_2\ldots dy_k = 0\end{aligned} \tag{5.33}$$

then

$$\int_\Omega f(y)dy = f_0. \tag{5.34}$$

As a result,

$$Z - E(Z) = f_0 + \sum_{i=1}^{k} f_i(y_i) + \sum_i \sum_{j>i} f_{ij}(y_i,y_j) + \ldots + f_{1,2,\ldots,k}(y_1,y_2,\ldots,y_k). \tag{5.35}$$

As terms are orthogonal, Eq. (5.35) can be squared and integrated over Ω, and decompose the variance of $f(y)$ into terms of increasing dimensionality as shown in Eq. (5.36):

$$\sigma_z^2 = \sum_{i=1}^{k} \sigma^2_{E(Z/y_i)} + \sum_i \sum_j \sigma^2_{E(Z/y_i,y_j)} + \ldots + \sigma^2_{E(Z/y_1,y_2,\ldots,y_k)}. \tag{5.36}$$

Now define sensitivity indices as:
$S_i = \sigma^2_{E(Z/y_i)} \big/ \sigma_z^2$: First order index. Corresponds to main effect of y_i

$S_{ij} = \sigma^2_{E(Z/y_i,y_j)} \big/ \sigma_z^2$: Second order index. It measures the effect of pure interaction between any pair of factors on the output. The remaining sensitivity indices apply the same terminology Then write as

$$1 = \sum_{i=1}^{k} S_i + \sum_i \sum_j S_{ij} + \ldots + S_{1,2,\ldots,k}. \tag{5.37}$$

Therefore, in this case, sensitivity of x with respect to R_f (first order index) is given by

$$S_{R_f} = \sigma^2_{E(x/R_f)} / \sigma_x^2. \tag{5.38}$$

Now, $\sigma^2_{E(x/R_f)}$ and σ^2_x are unknown parameters, which can be estimated from the pair of observations generated. σ^2_x is estimated by its unbiased estimator as shown in

$$S_x^2 = \frac{1}{(N-1)}\sum_i (x_i - \bar{x})^2, \tag{5.39}$$

which is also known as sample variance of *x*, and $\sigma^2_{E(x/R_f)}$ is equal to $\sigma^2_{g(R_f)}$, which can be estimated by its unbiased estimator in Eq. (5.40):

$$S^2_{E(x/R_f)} = \frac{1}{(N-1)}\sum_i (g_i - \bar{g})^2, \tag{5.40}$$

where $g_i = g(R_{f_i})$. Here, the true *g* is unknown but its estimate $\hat{g}$ can be obtained by the traditional nonparametric regression method. Hence, $\sigma^2_{E(x/R_f)}$ can be estimated by

$$S^2_{E(x/R_f)} = \frac{1}{(N-1)}\sum_i (\hat{g}_i - \bar{g})^2, \tag{5.41}$$

where $\hat{g}_i = g(R_{f_i})$.

Now, segment the range of R_f in some subranges also determine the sensitivity in those individual subranges.

Step 5 Determine optimal pair of (*x*, R_f) From the different sensitivity indices in different subranges of R_f, the sub-range of R_f that corresponds to least sensitivity can be chosen. The optimal pair of (*x*, R_f) should be the pair that corresponds to minimum of mismatch computed using Eq. (5.28).

5.4.2.3 Test Study

An actual utility case study is presented here. In the faulted network shown in Fig. 5.20, a DFR installed on bus 1 is triggered upon the occurrence of the fault and DFR report indicates the fault is on the line section 1–5 while actually the fault was reported to be 3 miles from bus 8. With the proposed algorithm, the fault is recognized as being either in line section 6–5 or in line section 7–9.

Fault resistance is chosen in the range of 0 per unit to 0.8 per unit, and fault resistance is changed within its range by increasing its value by 0.008 per unit in each iteration.

Fault resistance versus fault location is plotted in Fig. 5.21. Sensitivity analysis with respect to fault resistance is performed for several iteration runs and the sensi-

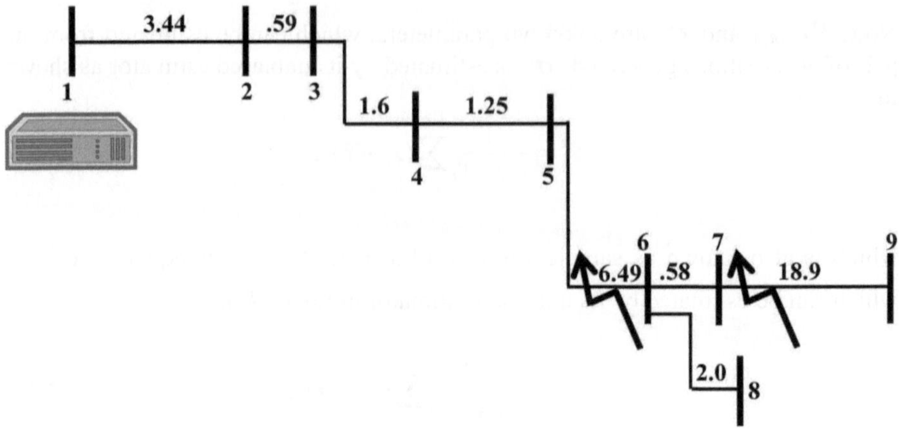

Fig. 5.20 Faulty network

Fig. 5.21 R_f versus x plot

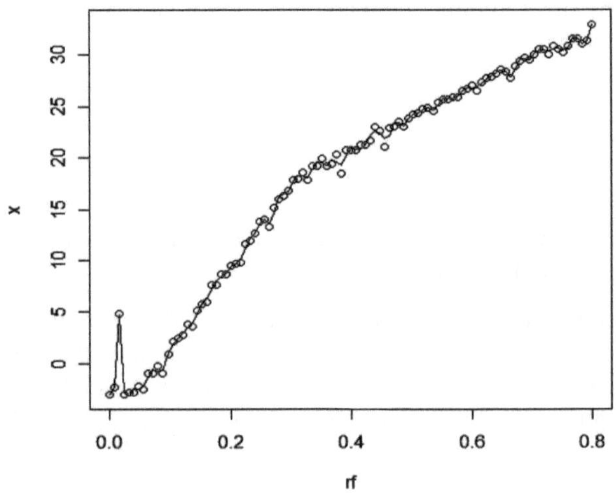

Table 5.3 Sensitivity to R_f for Different R_f Ranges

Sensitivity to $R_f\left(S_{R_F}\right)$	Range of R_f (per unit)							
	0–0.1	*0.1–0.2*	*0.2–0.3*	*0.3–0.4*	*0.4–0.5*	*0.5–0.6*	*0.6–0.7*	*0.7–0.8*
Run 1	0.04	0.98	0.99	0.65	0.71	0.95	0.99	0.99
Run 2	0.02	0.88	0.97	0.94	0.85	0.99	0.84	0.69

tivity indices corresponding to different subranges of R_f are presented in Table 5.3. The sensitivity indices reflect that for fault resistance in the range of 0–0.1 per unit, the sparse measurement-based fault location method yields accurate results, but after that the algorithm becomes sensitive to the choice of fault resistance.

5.4.3 Features and Benefits

A system-wide sparse measurement-based fault location approach using sparsely measured synchronized phasor is presented in this section. This method locates fault on a transmission line using recordings from different IEDs (DFRs in this example) triggered by a fault occurrence and makes use of SCADA data to update the power system model prior to the occurrence of the fault which makes the approach robust and accurate as well. The main advantages of this approach are:

- It is possible to calculate fault location accurately for the cases where other relay- or DFR-resident algorithms may fail due to the lack of measurements at the ends of the faulted transmission line.
- The fault location is robust since the same measurements may be obtained by multiple IEDs, which allows redundancy that can be explored to account for bad measurements.
- The computation may be totally automated allowing obtaining rather quick answer about the fault location even if performing the match may involve rather complex steps.

References

1. M. Kezunovic, Smart fault location for smart grids. IEEE Trans. Smart Grid **2**(1), 11–22 (2011)
2. IEEE Guide for Determining Fault Location on AC Transmission and Distribution Lines," IEEE Std C37.114–2004, 2005
3. M. Kezunovic, B. Perunicic, in *Wiley Encyclopedia of Electrical and Electronics Terminology*, vol. 7, ed. by J.G. Webster. Fault location (Wiley, New York, 1999), pp. 276–285 (ISBN 978-0-4713-4608-1)
4. B. Hoffmann-Wellenhof, H. Lichtenegger, J. Collins, *GPS: Theory and Practice*, 3rd edn. (Springer-Verlag, New York, 1994)
5. P. Dutta, A. Esmaeilian, M. Kezunovic, Transmission line fault analysis using synchronized sampling. IEEE Tran. Power Syst., in press
6. A. Gopalakrishnan, M. Kezunovic, S.M. McKenna, D.M. Hamai, Fault location using distributed parameter transmission line model. IEEE Trans. Power Deliv. **15**(4), 1169–1174 (2000)
7. M. Kezunovic, B. Perunicic, Automated transmission line fault analysis using synchronized sampling at two ends. IEEE Tran. Power Syst. **11**(1), 441–447 (1996)
8. M. Kezunovic, B. Perunicic, J. Mrkic, An accurate fault location algorithm using synchronized sampling. Elect. Power Syst. Res. **29**(3), 161–169 (1994)
9. M. Kezunovic, Optimized fault location. PSerc Report #08–07, April 2008
10. M. Kezunovic et al., Accurate fault location in transmission networks using modeling, simulation and limited field recorded data. PSerc Report # 02–44, Nov 2002
11. P. Dutta, Transmission line fault location using interoperability and integration of data and model. Ph. D. Dissertation, Texas, A&M University, May 2014
12. A. Gopalakrishnan, Fault location and parameter estimation on overhead transmission lines using synchronized sampling. Aug 2000

13. R.L. Eubank, *Nonparametric Regression and Spline Smoothing* (Marcel Dekker, Basel, 1988)
14. L. Collatz, *The Numerical Treatment of Differential Equations* (Springer-Verlag, New York, 1966)
15. S.M. McKenna, D.M. Hamai, M. Kezunovic, A. Gopalakrishnan, Transmission line modeling requirements for testing new fault location algorithms using digital simulators. *Proceeding of the second international conference on digital power system simulators*, Montreal, Quebec, May 1997
16. EPRI, Electromagnetic transients program (EMTP). *Revised Rule Book,* version 2, vol. 1, Palo Alto, CA, Mar. 1989
17. M. Kezunovic et al., Digital models of coupling capacitor voltage transformers for protective relay transient studies. IEEE Trans. Power Deliv. **7**(4), 1927–1935 (1992)
18. M. Kezunovic et al., Experimental evaluation of EMTP-based current transformer models for protective relay transient study. IEEE Trans. Power Deliv. **9**(1), 405–413 (1994)
19. A.O. Ibe, B.J. Cory, A traveling wave fault locator for two- and three-terminal networks. IEEE Trans. Power Deliv. **1**(2), 283–288 (1986)
20. J. Joe-Air et al., An adaptive PMU based fault detection/location technique for transmission lines—Part I: theory and algorithms. IEEE Trans. Power Deliv. **15**(2), 486–493 (2000)
21. J. Joe-Air et al., An adaptive PMU based fault detection/location technique for transmission lines—Part II: implementation and performance evaluation. IEEE Trans. Power Deliv. **15**(4), 1136–1146 (2000)
22. OSISoft, PI Historian. http://www.osisoft.com/
23. M. Kezunovic, Y. Liao, Fault location estimation based on matching the simulated and recorded waveforms using genetic algorithms. Development in Power System Protection, Amsterdam, The Netherlands, April 2001
24. M. Kezunovic, S. Luo, A novel method for transmission network fault location using genetic algorithms and sparse field recordings. *Proceeding of the IEEE/PES summer meeting,* July 2002
25. M. Kezunovic, S. Luo, Z. Galijasevic, D. Ristanovic, Accurate fault location in transmission networks using modeling, simulation and limited recorded data. PSerc Report #02–44, Nov 2002
26. S. Luo, M. Kezunovic, D.R. Sevcik, Locating faults in the transmission network using sparse field measurements, simulation data and genetic algorithms. Elect. Power Syst. Res. **71**(2), 169–177 (2004)
27. M. Kezunovic, P. Dutta, Fault location using sparse wide area measurement. *CIGRE study committee B5 annual meeting and colloquium,* Jeju Island, Korea, Oct 2009
28. P. Dutta, M. Kezunovic, Unified representation of data and model for sparse measurement based fault location. *IEEE/PES general meeting*, July 2012
29. D.E. Goldberg, Genetic algorithms in search, optimization and machine learning. Addison-Wesley, 1989
30. P. Dutta, M. Kezunovic, Fault resistance sensitivity of sparse measurement based transmission line fault location. *43rd North American power symposium (NAPS)*, Aug 2011
31. I.M. Sobol, Sensitivity analysis for non-linear mathematical models. Math. Model. Comput. Exp. 1 **1**(4), 407–414 (1993)

Index

M. Kezunovic et al., *Application of Time-Synchronized Measurements in Power System Transmission Networks,* Power Electronics and Power Systems,
DOI 10.1007/978-3-319-06218-1, © Springer International Publishing Switzerland 2014

MIX
Papier aus verantwortungsvollen Quellen
Paper from responsible sources
FSC® C105338

If you have any concerns about our products,
you can contact us on
ProductSafety@springernature.com

In case Publisher is established outside the EU,
the EU authorized representative is:
Springer Nature Customer Service Center GmbH
Europaplatz 3, 69115 Heidelberg, Germany

Printed by Libri Plureos GmbH
in Hamburg, Germany